A GEOGRAPHY OF
agriculture

THE BROWN
FOUNDATIONS OF GEOGRAPHY
SERIES

Consulting Editor
ROBERT H. FUSON
University of South Florida

A GEOGRAPHY OF

Agriculture
James R. Anderson, University of Florida
Transportation and Business Logistics
Edwin J. Becht, University of Oklahoma
Plants and Animals
David J. de Laubenfels, Syracuse University
Geography
Robert H. Fuson, University of South Florida
The Atmosphere
John J. Hidore, Indiana University
Population and Settlement
Maurice E. McGaugh, Central Michigan University
Industrial Location
E. Willard Miller, The Pennsylvania State University
Water
Ralph E. Olson, University of Oklahoma
Earth Form
Stuart C. Rothwell, University of South Florida
Minerals
Walter H. Voskuil, University of Nevada

THE BROWN
FOUNDATIONS OF GEOGRAPHY
SERIES

A GEOGRAPHY OF

agriculture

JAMES R. ANDERSON

University of Florida

WM. C. BROWN COMPANY PUBLISHERS
DUBUQUE, IOWA

THE BROWN
FOUNDATIONS OF GEOGRAPHY
SERIES

Consulting Editor
ROBERT H. FUSON
University of South Florida

Copyright © 1970 by
Wm. C. Brown Company Publishers

ISBN 0–697–05150–1

Library of Congress Catalog Card Number: 76-112619

Second Printing, 1971

Printed in the United States of America.

Geography is one of man's oldest sciences, yet it is as new as the Space Age. Knowledge of the earth obtained from satellite photography and measurement, remote sensing of the environment, and by means of other sophisticated techniques are really but a stage in the evolutionary process that began with ancient man's curiosity about his surroundings. Man has always been interested in the earth and the things on it. Today this interest may be channeled through the discipline of geography, which offers one means of organizing a vast amount of physical and cultural information.

The **Brown Foundations of Geography Series** has been created to facilitate the study of physical, cultural, and methodological geography at the college level. The **Series** is a carefully selected group of titles that covers the wide spectrum of basic geography. While the individual titles are self-contained, collectively they comprise a modern synthesis of major geographical principles. The underlying theme of each book is to foster an awareness of geography as an imaginative, evolving science.

Preface

One of the most persistent and compelling problems of today's world is that of hunger and poverty.

This book is written for students taking introductory level courses in geography in order that these students may more fully appreciate the significance of agriculture as a primary economic activity in resolving this basic problem. The first chapter relates agriculture to the need for food and fiber. The role of agriculture as a vital force in world affairs is given proper perspective by contrasting two worlds that are now trying to exist together on the planet earth. These worlds are simply referred to as the World of Plenty and the World of Poverty. In the second chapter, agriculture as an economic activity is related to the land resource base and to the application of technology in using these resources more efficiently. The pattern of world agriculture, systems or types of agriculture, and recent trends in agricultural production are other topics discussed. Within the prospectus for this book it has not been possible to deal with the historical facets of the agricultural landscape and its evolution and with the origin and dissemination of crop varieties and animal breeds throughout the world over the long span of time that man has existed on this earth.

Obviously the book is addressed primarily to students of geography in the United States and Canada. The author has generally used examples from the United States, since it is hoped that some familiarity of students with the agricultural activities of the country in which they live will help them appreciate the areal differences that exist in the character and volume of agricultural activity.

The author's eight years of prior association with the U.S. Department of Agriculture has enabled him to utilize current research being carried on in that department, particularly by the several divisions of the Economic Research Service.

From early youth when the author was growing up on a farm in south-central Indiana, through the years of university life while studying under such men as Oliver E. Baker, L. Dudley Stamp, C. Warren Thornthwaite, William D. Thornbury, and Stephen S. Visher, his interest in agriculture has been a dominant one. To the many students at the Universities of Maryland, Virginia, and Florida who have been in his classes, the author owes much for their inspiration and enthusiasm. To his wife, Joy, the author is especially indebted for her understanding support of his work on this book.

James R. Anderson

Contents

Agriculture and the Need for Food and Fiber

Food and fiber are vital to the existence of man in the world in which he lives. When the production of these vital commodities is examined geographically, it is obvious that great contrast exists in the availability of these basic necessities for the world's people. In reality there are two worlds today—A World of Poverty and A World of Plenty.

With the many political, social, and economic crises that exist today, poverty and hunger are generally not accorded the publicity that such fundamental problems deserve. The fact that about three-fifths of the world's people with inadequate diets are really living in a world apart is widely overlooked. This world will be called the World of Poverty. It is a world not only characterized by hunger but by deprivation that denies the people living in it many of the other basic necessities of a decent way of life to say nothing of their lack of the material goods and the leisure time that provide the many comforts beyond life's basic needs. China, India, most of southeastern and southwestern Asia, much of Africa, several countries of Latin America, and part of eastern and southern Europe comprise this world.

The other world consists of that more fortunate two-fifths of the world's population living in the United States, Canada, northwestern Europe, Australia, New Zealand, Italy, Poland, Czechoslovakia, U.S.S.R., Japan, Argentina, Chile, Uruguay, Venezuela, the Republic of South Africa, and perhaps a few smaller countries such as Israel. This world will be called the World of Plenty. In this world the relative abundance of resources in relation to population density and technological innovation have permitted a level of living that is considerably above a minimal survival level.

Within both the world of poverty and the world of relative plenty, considerable variation exists in such indicators of the level of economic

development as per capita gross domestic income, per capita consumption of inanimate energy, percent of the labor force in agriculture,
and dietary levels. In spite of the excellent communications networks
which are capable of keeping man in touch with the activities and
problems of both worlds, there is a lack of understanding and a general
lack of concern that is in part attributable to ineffective education about
the physical, economic, social, political, and other geographic characteristics of the two worlds of the present time.

The role of agriculture in resolving the basic problem of producing
more food and fiber for more people has a geographical perspective.
The following facets of the basic problem must appropriately be examined as a background for a geographical analysis of agricultural production.

1. Progressive but unequal acceleration of population growth
2. Uneven distribution of population
3. Relation of population density to technological levels
4. Variations in dietary levels and composition of the diet

Population Growth

World population has increased persistently at an accelerating rate
for at least 300 years. Between 1650 and 1850, a period of 200 years,
the estimated population of the world approximately doubled. Between
1850 and 1950 the population of the world doubled again, this time in
100 years. Obviously the present rates of growth clearly indicate that
the world population is now doubling again, this time in less than
50 years. This acceleration of population growth is generally outstripping gain in food production in the World of Poverty. Thomas Malthus
stated in 1798 that the "power of population is indefinitely greater
than the power of the earth to produce subsistence for man."[1] While
the Malthusian statement on population growth no longer is valid for
the World of Plenty, it still has relevance when evaluating conditions
in the World of Poverty. The per capita world food supply is currently
increasing about one percent per year while the world's population is
growing about two percent per year.

Startling though the difference between world population growth and
the increase in world food supply may be, it is the differential between
the World of Plenty and the World of Poverty that is especially a
matter of concern. In the early decades of the present century the
rates of population growth for Asia, Africa, and Latin America actually

---

[1]Thomas R. Malthus, *An Essay on the Principle of Population*, 2nd edition, New
York: The Macmillan Company, 1929, p. 68.

lagged behind those of the economically developed parts of the world. But with medical advances and improvements in health and sanitation the less developed parts of the world have had very rapid increases in population in recent decades. The population of Latin America has grown most rapidly but many of the densely populated areas of Asia which had a much greater population base at the beginning of the century currently have the most critical problems.

Any evaluation of the present crisis in world population growth must consider carefully the present impact and future possibilities of population control in finding a satisfactory answer to fundamental questions. How can a level of living approaching that now found in the World of Plenty be achieved for the World of Poverty? Some would also ask: Is it really within the realm of possibility to achieve such an end?

Long overlooked has been the role of population control in the foreign aid and domestic poverty programs of the United States. It has only been within the present decade that a basic change in policy has begun to evolve in this country. There is nothing new about population control. Much evidence exists indicating that some measures that have long been used are celibacy, continence, late marriage, infanticide, cannibalism, and abortion to control population growth in past societal situations where economic circumstances were unfavorable to unlimited increases in population.

In Ireland the average age of marriage has risen to 28, and more than a fourth of Irish women remain unmarried at age 45. In the United States today the average age of marriage for the woman is about 20 years, but seven to nine million American women are now using the pill. Differing attitudes on the sensitive issue of birth control are very much a part of the massive problem of accelerating population growth.

Changes in attitude toward the use of birth control measures are occurring in this country. Shortly before his death, a recording of the voice of former President Dwight D. Eisenhower was broadcast on a paid radio advertisement in which the former President stated unequivocally that he had changed his mind about the need for birth control in arresting excessive growth of population. The late President Kennedy was the first President to take a definitive position in favor of spending foreign aid money on family planning. Five dollars spent abroad on family planning will probably have more impact than $100 spent on programs to increase food production.

Generally it has not been widely recognized that the religious beliefs of most of the people of Asia are not necessarily a barrier to the acceptance of the idea of limiting the size of the family. Strong emphasis on the perpetuation of the family and the desire for male heirs in countries such as China does exist; but as Professor Charles Hu once remarked in his seminar on South and East Asia: "A second, third, or fourth son is an insurance policy. If a Chinese father had a reasonable assurance that one son would live to rear a family, the need for the high insurance costs of the additional sons would disappear."

Population Distribution

A particularly geographical facet of the dilemma of two worlds is the distribution of world population. When one looks at a map showing the distribution of population over the world, he is immediately impressed by the great contrasts that exist. Vast areas in polar regions, extensive deserts, and several million square miles of the wet tropics have very sparse populations. On the other hand, one can point to parts of the world where extremely high concentrations of population do exist. Southern and eastern Asia and western Europe are especially noticeable. Northeastern United States, although lesser in extent, is also becoming an area of significant concentration. A glance at the breakdown of population by major world regions shown in TABLE 1.1 indicates clearly the dominance of Asia (excluding the U.S.S.R.) in number of people. This massive concentration on a relatively small part of the world land resource base is a staggering situation.

TABLE 1.1
World Population by Major World Regions, 1650-1968
Number in Millions

	1650	1750	1850	1950	1968
Europe*	100	140	266	394	455
Anglo America	1	2	26	168	222
Latin America	12	11	33	163	268
Asia*	330	479	749	1,380	1,943
Africa	100	95	95	199	333
Oceania	2	2	2	13	19
U.S.S.R.	N.A.	N.A.	N.A.	180	239
Total	545	729	1,171	2,497	3,479

*Excludes U.S.S.R. in 1950 and 1968
N.A.—Not available

Source: L. Dudley Stamp, *Land for Tomorrow*, Bloomington: Indiana University Press, 1952. Data for 1950 are from *The Future Growth of World Population*, New York: United Nations, 1958. Data for 1968 from Population Reference Bureau, Washington, D. C.

When the distribution of world population is compared with the distribution of the world's cropland, which contributes from two-thirds to three-fourths of the world's food, it is obvious that the two distributions are very similar in the World of Poverty. In the World of Plenty, extensive areas of cropland are not so closely related to some of the major concentrations of population, particularly where marked urbanization has occurred. The major cropland area of Canada and the cropland concentration in the Great Plains of the United States are striking examples of fertile cropland areas that have relative sparse populations.

Some striking differences in the distribution of the present cropland area in relation to the distribution of land that may be suitable for future crop production may of course be observed and this important matter will be discussed in Chapter 2. The possibilities of expanding cropland in the wet tropics and of using desalinization of sea water as a means of increasing cropland at least on the fringes of deserts offer distinct possibilities for some marked expansion of the cropland base.

TABLE 1.2

Current Annual Rate of Population Growth and
Crude Birth and Death Rates by Major World Regions, 1968

	Percent	Birth Rate‡	Death Rate‡
Europe*	0.7	18	10
Anglo America†	1.1	19	9
Latin America	3.0	40	10
Asia*	2.2	39	17
Africa	2.3	45	22
Oceania	1.8	20	9
U.S.S.R.	1.1	18	7
World	2.0	35	14

*Excluding U.S.S.R.
†Canada and United States
‡Per 1,000 population

Source: People: An Introduction to the Study of Population, Population Reference Bureau, Washington, D. C., 1968.

Yet, these very possibilities, exciting though they may be from a technological standpoint, are often situations that have the greatest institutional barriers to effective use for raising levels of living in the World of Poverty. Migration of population from areas of mass poverty to areas offering opportunities for a better level of living has generally been beset by political, social, and economic barriers that are almost insumountable. Only a passing reference to the immigration laws of the United States or to those of Australia need be made to emphasize this roadblock to a solution.

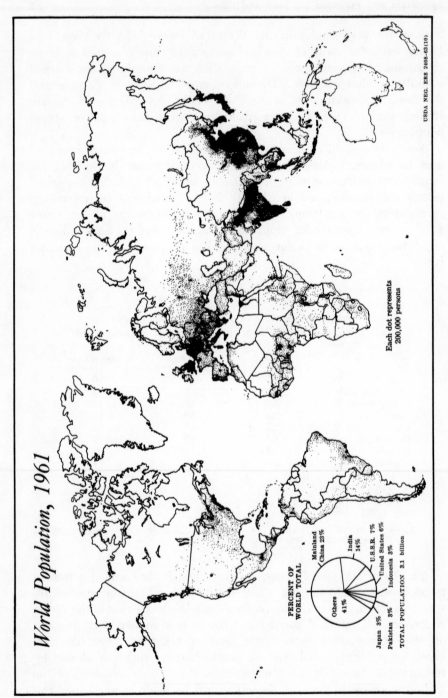

World Population, 1961

PERCENT OF
WORLD TOTAL

Mainland
China 23%

India
14%

U.S.S.R. 7%

United States 6%

Indonesia 3%

Others
41%

Japan 3%

Pakistan 3%

TOTAL POPULATION 3.1 billion

Each dot represents
200,000 persons

USDA NEG. ERS 2408–63(10)

FIGURE 1.1.

Source: A Graphic Summary of World Agriculture, p. 11, U.S. Department of Agriculture. 1964.

6

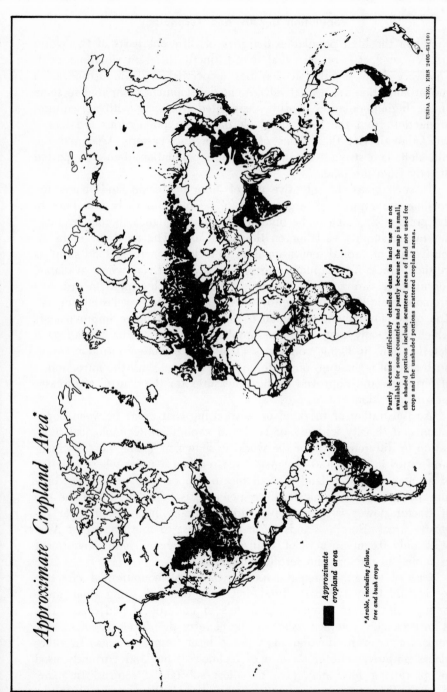

Approximate Cropland Area*

*Arable, including fallow, tree and bush crops

▮ Approximate cropland area

Partly because sufficiently detailed data on land use are not available for some countries and partly because the map is small, the shaded portions include scattered areas of land not used for crops and the unshaded portions scattered cropland areas.

USDA NEG. ERS 2405-63(10)

FIGURE 1.2.

Source: A Graphic Summary of World Agriculture, p. 8, U.S. Department of Agriculture, 1964.

7

Population Density and Technology

When the high population densities of different parts of the world are examined, it is obvious that two distinctly different situations exist. The very high population densities associated with industrialization and urbanization are a relatively recent development compared to some of the high population densities associated primarily with agricultural production from the area in which the dense populations are found. Java, the alluvial floodplains of southern and eastern Asia, and the Nile delta are striking examples of dense populations being supported directly from the land.

In such areas the intensive use of labor on fertile lands has made possible the support of large masses of people. But there is a limit to the production that can be obtained from these lands merely by making further inputs of labor on the same land. The need for more intensive use of capital seems apparent in order to increase food supplies in such areas. Yet capital is extremely scarce. Even when available, considerable care must be taken in investing such capital in order to promote basic long-range improvements that will have a lasting impact. Too often, investment of capital is made in short range improvements which may only aggravate the problem and may actually lead to a deterioration in living conditions. Thus the effective introduction of birth control techniques ought to go hand in hand with the introduction of fertilizers and improved medical measures for the control of diseases and malnutrition.

An illustration of the kind of relationship that exists between population and density and use of land and capital for technological innovation in different parts of the world is shown in TABLE 1.3. The man-land ratios in south and east Asia have reached a point where only 0.6 of an acre of arable land is available to support one person, yet only nine pounds of inorganic fertilizer per acre is being used and the use of tractor power is very uneven and generally non-existent over extensive areas. Europe stands in sharp contrast to south and east Asia. With only 0.8 of an acre of arable land per person, much investment of capital in tractors and fertilizer has occurred.

One of the fundamental issues in the implementation of effective foreign aid is to use scarce funds in a way that will bring about more lasting solutions to economic problems in underdeveloped areas. This is an extremely complex issue. In one country the development of new lands for crop production may bring basic improvements. In other places, improvement in the use of technology on land currently used for producing food may give the most substantial returns. In many instances these two approaches might be used simultaneously. Certainly, one cannot overlook the emphatic point that concurrently the matter

TABLE 1.3
*Arable Land Per Person and Tractors and Fertilizer
Use Related to Arable Land by Major World Regions, 1965*

	Acres of Arable Land (Cropland) Per Person	Acres of Arable Land (Cropland) Per Tractor	Pounds of Nitrogen, Phosphate and Potash Applied Per Acre of Arable Land
Oceania	5.2	227	36
Anglo America*	2.6	108	46
U.S.S.R.	2.5	343	22
Latin America	2.2	957	6
Africa	1.8	2,185	4
Europe†	0.8	75	106
Asia†	0.6	3,583	9‡

*United States and Canada
†Excluding U.S.S.R.
‡Excluding Mainland China

Source: Based on data from the *Production Yearbook*, Food and Agriculture Organization of the United Nations, Volume 20, 1966.

of population control must be faced realistically and squarely if lasting solutions are to be found.

Granted that a workable approach can be developed and assuming that funds are available for program effectuation, one must still come back to the recognition of the fact that it is the areas of extremely high population densities that have the least probability of effective solutions in a current lifetime.

Dietary Quality

Dietary levels of different parts of the world contrast greatly. Some of this contrast is attributable to differences in requirements, which vary with climate (more calories are needed in cold climates than in warm climates), with average body size, and with the level of an individual's physical activity. Daily per capita nutritional reference standards vary from 2,710 calories required in such countries as Canada and the U.S.S.R. to less than 2,400 calories required for an adequate diet in areas such as Africa and Mainland China. Contrasts in the fat requirements for an adequate diet also vary particularly with climate. Protein, the third major component of the diet, is needed in similar amounts from place to place by persons of the same age.

In meeting the caloric needs of the world's population, rice and wheat are the dominant sources of food energy. These two grains along with corn, millet, and sorghum accounted for just about one-half of man's food energy in 1958. Striking regional contrasts in the sources of calories exist from one part of the world to another. In Asia excluding the U.S.S.R., nearly three-fourths of all calories are derived from grain products, roots, and tubers, whereas in the United States and Canada only about a fourth of the calories come from this source. In Oceania (particularly in Australia and New Zealand) as well as in the United States, animal products constitute a very important source of calories (Figure 1.3).

The level of income available to a given population is another major determinant of the dietary level. Using the nutritional reference standards as a basis for comparison, it is obvious that most of the people of the World of Plenty are eating too much while those from the World of Poverty are of course undernourished (Figure 1.4). When the situation is analyzed carefully, it soon becomes evident that the problem is a very complex one to which there is no easy or single solution. Yet obviously a higher level of food production is needed in many countries. A better distribution of food presently produced would also help to alleviate some problems if the economics of redistribution could be satisfactorily resolved. A change in dietary habits away from traditional patterns would also contribute to improved nutrition and in some places

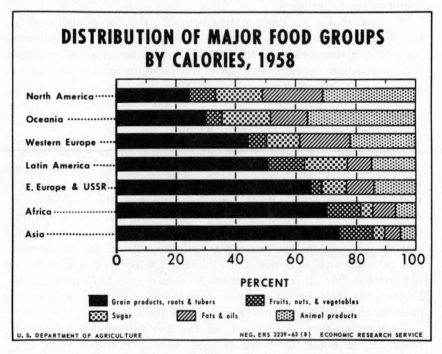

FIGURE 1.3.

Source: L. R. Brown, Man, Land, and Food: Looking Ahead to World Food Needs, p. 28. U.S. Department of Agriculture, 1963.

could be accomplished without needing any major change in the total level of agricultural production.

Undoubtedly, changes in dietary habits are closely associated with level of education and of course with levels of income if there is an increased cost associated with a change in diet. The growing consciousness of the American people about dietary matters has in part coincided with the exposure of millions of persons to military diets during and following World War II as well as restrictions placed on the dietary habits of the civilian population by rationing of food during the War. The research of many scientists such as Professor Ancel Keys, a physiologist at the University of Minnesota and the developer of K rations (K for "Keys"), was also a major influence in bringing about dietary changes. A number of non-fiction "best sellers" offered advice on how to lose weight quickly but safely. The impact of scientific research such as that carried out by Keys and others is much greater in countries where people are well-educated and are exposed to mass communications media such as the newspaper, radio, and television.

The research of Keys and others also has interesting implications for the dilemma of the two worlds—the one world with too little food and the other with too much. When Professor Keys proposed an "ideal" diet for Americans some years ago, he emphatically pointed out that Americans, who have become mainly a people who work sitting down, eat too much. He would promptly reduce the average American from a diet of 3,200 calories per day to 2,300 calories. More significantly, he would make a major change in the composition of that diet. The average American diet is composed of about 46 percent carbohydrates, 14 percent proteins, 23 percent unsaturated fats (vegetable oils, fish, etc.), and 17 percent saturated fats (animal meat, eggs, and dairy products). The proposed diet by Keys was composed of 69 percent carbohydrates (spaghetti is a good basic food according to Keys), 16 percent proteins, 11 percent unsaturated fats, and 4 percent saturated fats. Quite a significant recommendation really!

It must be recognized that nutritionists are not in full agreement on what constitutes an ideal diet. Particularly troublesome is the problem of proteins. Proteins derived from plant sources are more likely to be lacking in some of the essential amino acids, which are more readily available in animal proteins. Yet, if one assumes that the diet of Professor Keys is basically sound, this is an important premise on which to plan for basic improvements in world dietary conditions. Furthermore we have a diet which will be far more easily supplied than the all-meat diet which the late Dr. Vilhjalmur Stefansson, under whom

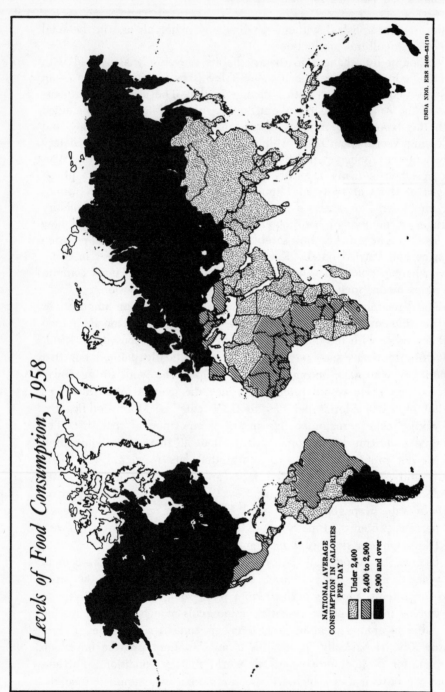

Levels of Food Consumption, 1958

NATIONAL AVERAGE
CONSUMPTION IN CALORIES
PER DAY

Under 2,400

2,400 to 2,900

2,900 and over

USDA NEG. ERS 2460-63(10)

FIGURE 1.4.

Source: *A Graphic Summary of World Agriculture*, p. 12, U.S. Department of Agriculture, 1964.

the author studied at McGill University, insisted was the natural food of man.

The Chinese are not vegetarians by choice but by necessity. The author remembers Professor Hu referring to the chicken as the vacuum cleaner and the hog as the garbage can of the Chinese home. Cereals and other crops consumed directly by people supply far more calories than when these products are first fed to livestock before human consumption takes place. For example, corn will yield eight to ten times as many calories if consumed directly rather than if first converted into beef or pork.

Thus, if a satisfactory diet with a higher proportion of cereals, legumes, fruits, vegetables, and fish can be developed to supply the necessary calories, proteins, and vitamins and if dietary traditions and customs can be changed, then there may be a chance to improve the diets in the World of Poverty.

In Summary—A Solution?

In summary, can some significant answers to some penetrating questions be found? Can any conclusions be reached? Is there a solution? Can the World of Plenty and the World of Poverty continue to exist side by side? Can science open up the necessary advances in technology to bring about improvements? Are new lands available and satisfactory for the extension of agricultural production to this previously little-used land? Can foreign aid by the governments of the economically developed countries be made more effective? Will such aid programs really make a sufficient impact to get the job done? These are complex questions that certainly have no easy and ready answers. Yet, answers to these must be found soon.

It must be fully realized that the problems outlined above must be attacked on a number of fronts simultaneously. There is a need to continue to strive for advances and breakthroughs that will lead to substantial increases in food production. A better distribution of food supplies, now a major economic bottleneck, must be found. The search for ways to increase food supplies must be diverse and effectively fitted to many different situations. Intensification of effort on the existing cropland base and extension of the cropland area, and a combination of technological innovation and the extension of agriculture into areas not previously used will need to be implemented after careful planning.

New lands for the extension of crop production are definitely available and in sufficient quantity to make a major impact on world food production. Of course, it will cost more to put these lands into production but this is to be expected. A recent report, *Possibilities of Increasing World Food Production* by the Food and Agriculture Organization, concludes that the "potentials for increasing food production are very substantial indeed." Charles Kellogg, Chief of U. S. Soil Survey, also is quite optimistic about the availability of new lands for crop production, particularly in the wet tropics. His position on this matter has changed significantly in recent years, since more information is now available about some of the major world soils. This leads him to conclude that agricultural soils are available and can be used for various crops that need to be grown, particularly rice and other tropical and subtropical crops.

The possibilities of population control are even more challenging and the need for immediate action is even more critical. If swift and deft action is not taken to check population growth in areas where birth rates are now very high in relation to the capabilities of resources to feed existing populations, not much progress will be made toward raising levels of living for these areas. Certainly the trends in the U.S. birth rate offer evidence that economic and social factors do bring about a marked and rapid change in a few years. The decade of the fifties was a decade of rapid population increase in this country, and it has now been followed by a sharp decline in the birth rate which reached an all-time low of 17.9 births for every 1,000 Americans in 1967. The previous low of 18.4 occurred during the depression years of 1933 and 1936. The reasons for this decline are complex and perhaps are not even yet fully understood. Certainly the increase in the use of contraceptives has made a significant impact. Yet, the decline in the American birth rate must not be entirely attributed to the recent availability of a new type of contraceptive. As Phillip M. Hauser, Director of the University of Chicago Population Research Center, has stated, "the decision of couples to forego a third and fourth child, substituting perhaps, a second car and color TV" is a major factor accounting for the substantial decline in births.

The Japanese experience is a most significant example of rapid change in national trends in population growth, which has been brought about through specific and drastic policies that have had an immediate demographic impact. In 1952 the Eugenic Protection Law was revised to simplify procedures for legal induced abortion. A significant part of this revision provided for the legalization of abortion by licensed physicians with the consent of both spouses "if the continuance of pregnancy or the delivery seems remarkably injurious to the health of the mother due to her physical or *financial condition.*"

In 1947 the crude birth rate in Japan was 34.3 per thousand. By 1957, just ten years later, the crude birth rate had dropped to 17.2

per thousand. In 1949, 246,000 abortions were recorded in Japan. By 1957 this figure had reached 1,122,000. Voluntary sterilization was also markedly on the increase particularly in families where at least two children had already been born.

The impact of this Eugenic Protection Law on the population growth of Japan has been remarkable. Whether or not the low rate of growth continues into the future will in part be dependent upon the state of the Japanese economy and the influence of rising levels of living upon family planning. This is not really the point at this time. It is of major significance that such a marked change was accomplished in a short time. There are many who would not endorse the Japanese approach to population control. Yet it must be pointed out that it was an acceptable method of raising levels of living in a country which clearly recognized after World War II that its resources were quite inadequate to take care of marked increases in population growth if levels of living were to be improved. It is this kind of action that must be recognized as being vital and necessary in order to make a significant impact on current basic problems. The dilemma of the two worlds will not be solved unless there is drastic action. As one scientist recently put it, the world must choose between "high mortality and low fertility."

And Professor Kenneth Boulding so aptly stated: "What has posterity done for me? Some of us perhaps do think that posterity should have a voice even though it does not have a vote. I do believe that the society that loses its identity with the future also loses its capacity to deal with the present."

References

BROWN, HARRISON. *The Challenge of Man's Future,* New York: The Viking Press, Inc., 1954.

HOLLINS, ELIZABETH JAY (editor). *Peace is Possible,* New York: Grossman Publishers, Inc., 1966.

MUDD, STUART (editor). *The Population Crisis and the Use of World Resources,* Bloomington: Indiana University Press, 1964.

National Academy of Sciences Committee on Science and Public Policy. *The Growth of World Population,* Washington, D. C.: National Academy of Sciences, 1963.

ZELINSKY, WILBUR. A *Prologue to Population Geography,* Englewood Cliffs, N. J.: Prentice-Hall, Inc., 1966.

Chapter 2

Resources for Agriculture

Agriculture can be carried on almost anywhere. Bananas are produced in Iceland. However, from the standpoint of production costs it is hardly practical to grow a tropical crop in a polar region. In recent years man has made great strides in developing agricultural technology which now makes land resources formerly considered unsuitable for agriculture more usable for the production of crop and livestock products. The massive substitution of capital for labor and land has been a major force in bringing about substantial increases in agricultural output in the economically more developed countries. In those parts of the world where technological innovation is taking place at a slower pace, agricultural activities are more likely to be utilizing only land resources most suitable for such activities unless conditions of land ownership and other institutional restraints interfere.

The Resource Framework

Many years ago, O. E. Baker selected the world's most widely grown crop, wheat, and, after consulting with agronomists concerning the physical limits of its production, made some careful estimates of the area of the earth's land surface which was suited and unsuited to the production of this crop. Shortly before his death in 1949, Dr. Baker restudied the distribution of world wheat production and concluded at that time with the varieties and technology then used that 90 percent of the 52 million square miles of land area (excluding polar ice caps) was unsuitable for wheat production for the following reasons:

40 percent was too dry
21 percent was too wet (but much of this was suitable
    for rice)

21 percent was too cold (but some of this could be
used for forage crops)
6 percent was too rough
2 percent had unsuitable soils

Twenty years later, these limits of wheat production have not changed appreciably in spite of new wheat varieties developed in Mexico
and now being widely accepted by farmers in India and Pakistan for
use in areas which have long produced wheat. Similarly, the new hybrid
variety of rice, IR-8, developed by scientists sponsored by the Ford
and Rockefeller foundations in the Philippines has significantly increased yields but primarily on the same kind of land previously used
for rice culture.

What are the physical attributes of land that are favorable for the
production of agricultural commodities? Although the physical requirements of crops vary widely, there are some general attributes or qual-
ities which land must have to be economically usable for agricultural
production. The level of technological enhancement is also a very important determinant in the use of land for crop and livestock production.

Aridity is the most widespread limitation of crop production. In
general, land receiving less than ten inches of precipitation in the middle and high latitudes and less than 20 inches in the tropics is too
dry for crops unless irrigation is employed. Since antiquity, arid lands,
such as those of the Nile Valley and Mesopotamia, have been used
for agriculture by artificially applying water to the land. Water to
irrigate arid lands has generally been taken from exotic streams or from
reservoirs where the water from such streams is impounded. These
streams originate in areas of higher rainfall often far removed from
the alluvial plains that are being irrigated. In many places irrigated
land has provided the main base for a highly developed agricultural
economy. California is the leading agricultural state of the United States,
an eminence attained mainly by use of irrigated land and by concen-
trating particularly on the production of high value crops such as fruits
and vegetables.

Producing agricultural commodities in arid regions has many advantages. Among these are the relative freedom from insects and diseases and nearly the complete regulation of water supply. In humid
areas where irrigation of crops is being undertaken, a farmer may put
two inches of water on his crop and, as one farm manager in the
Mississippi Delta recently told the author, the next day "the Good Lord
gives us another two inches."

Yet the fact remains that availability of water at a cost which farmers
can afford to pay is the main determinant of the feasibility of using
arid areas for agriculture. In 1964, 37 million acres of land was irri-

gated in the United States. Of this irrigated land, 33 million acres were located in the arid West and about four million acres in the humid East. In the eleven western states, not including Hawaii and Alaska, there are 753 million acres of land nearly all of which would need irrigation if otherwise suited for agriculture. Only three percent of this land is presently being irrigated and additional water might be available to irrigate another three or four percent of the total area with better water management and if industrial and urban uses of water are not expanded greatly, a condition not likely to prevail.

Need for water for agricultural and other uses over vast expanses of arid land has raised the question in recent years about the feasibility of desalting sea water in order to obtain fresh water for densely populated arid lands, particularly where population growth is outstripping available water supplies. In some parts of the world, sea water is already being used for supplying urban centers with water. In southern California, plans for extensive supplementation of existing water supplies by desalinization of sea water are now being implemented. The cost of obtaining fresh water from the sea is still too high for much use to be made of such water in producing agricultural commodities, even in coastal areas. When costs of transporting desalted water to agricultural areas situated some distance from the sea are added to the costs of desalinization, then the costs are prohibitive in relation to the present prices being paid for agricultural products. The time may not be far off, however, when sea water will be used to irrigate crops of high value that are being grown in favorable locations near to large urban markets.

In spite of the great expanses of arid land that have little immediate potential for agricultural production, there are at the present time about 370 million acres of land being irrigated. Countries of southern and eastern Asia are using millions of acres of irrigated land for rice culture in areas having adequate moisture for the production of many other crops that can be grown without irrigation. The United States, U.S.S.R., Iran, Iraq, United Arab Republic (Egypt), Mexico, and Spain are countries with large areas of irrigated land.

Turning to coldness which is another major deterrent for agricultural activities, there is about a fifth of the world's land area which is too cold for the production of crops. Although barley and oats can be grown where the summers are a little shorter and a little cooler than those required for producing wheat, there is really very little production of any small grain or cereal crops (wheat, rye, oats, or barley) north of the 57° isotherm for the three summer months of June,

July, and August.[1] Some hay and forage crops are produced in areas
having lower summer temperatures and short frost-free seasons.

Most of the cold land areas, other than Antarctica which is permanently covered with ice, are located in the northern hemisphere,
since Africa, Australia, and South America do not extend into the high
latitudes. The U.S.S.R. and Canada have the largest expanse of land
too cold for crop production. Alaska, which has 365 million acres of
land area, has less than 25,000 acres of cropland. Sweden, Finland,
Norway, and Iceland also have areas too cold for much agricultural
activity.

In Canada some agricultural settlement in the Peace River Valley
has pushed the frontier of crop production well north of the main
areas of agricultural production in that country. In Alaska some experimentation with crops and livestock has been conducted and a few
farms may be found in the vicinity of Fairbanks. In the U.S.S.R. more
attention has been given to experimentation and the possibilities of
producing agricultural commodities in the cold northern part of that
country have been weighed carefully with other alternatives for increasing agricultural output. Generally more attention is being given
to the other alternatives, although the Soviets have given much publicity to some of their efforts to improve production possibilities for
barley, oats, hay and forage crops, and leafy vegetables such as cabbage in the extensive areas of that country having a short, cool summer.

The wet tropics, which are too humid for wheat production, are of
course well-suited to the production of rice, which is the world's second
most important basic food crop. Other crops also may be grown in the
wet tropics; yet today vast areas of land in this part of the world
remain unused for agricultural production. Of all the conditions which
are presently limiting crop production, the wetness and infertile soils
of the tropics offer one of the most immediate opportunities for extension of crop production into areas formerly considered unsuited
for such use. Formidable obstacles, many of them institutional (social,
political, and economic) rather than physical in nature are still to be
surmounted before much real progress in the agricultural development of new lands in the wet tropics can be accomplished. The vast
Amazon basin of Brazil and the eastern lowlands of Bolivia, Peru,
Ecuador, and Colombia, the African wet tropics, and more limited
areas in Southeast Asia are receiving more and more attention as areas
where new agricultural lands may be developed.

In addition to the climatic limitations of dryness, coldness, and
wetness which have been discussed, topography or the lay of the land

---

[1]An isotherm is a line connecting points having the same temperature.

is also of considerable significance. In some parts of the world having dense populations and where level arable land is scarce, much use is made of steeply sloping land for growing crops. In China, Japan, the Philippines, and other parts of south and east Asia the hill and mountain slopes have been laboriously terraced in order to produce crops without losing the valuable soil, which has often been greatly improved over many years of intensive management. In parts of Latin America and particularly in southern Europe, fields of crops may be seen on rough terrain. Vineyards, olive groves, and fruit crops are particularly important on hill slopes in some areas of Europe. In parts of the United States, particularly in the northeastern hill lands, apples and peaches are often grown on steeply sloping land. This type of land would hardly be used for other crops any longer in this country except in the Appalachians and the Ozarks where steep slopes are still utilized, often with rapid loss of the thin layer of top soil, on small farms which have very little level land available for crop production.

Generally the rough land of the United States, much more of which was formerly used for agriculture, is no longer economically usable since the substitution of the tractor for the horse and mule. Horses and mules could be used more easily and efficiently in small fields on steep slopes than the tractor-drawn equipment which has so rapidly replaced that drawn by animals in this country.

Soils of many different kinds can be and are used for producing crops. Some soils are better suited for some crops than for others. Some soils are inherently or naturally very fertile; others through intensive management are made more fertile than they originally were when first used by man. Because of favorable temperature and moisture conditions some soils are more productive but less fertile than others. For example, the chernozems or the black soils of the Dakotas are among the world's most fertile soils; yet the prairie soils of Illinois and Iowa are more productive because of the more favorable temperature and precipitation conditions.

Only a very small proportion of the world's soils are completely unmanageable. Some of the laterites of the tropics are apparently not economically amendable with techniques now available, and highly alkaline soils of the deserts are very difficult to use for crop production. Over extensive tundra areas of the high northern latitudes, very little soil formation has taken place which, combined with the severe climate, adds another major handicap to the agricultural utilization of such areas.

Present Use of Land for Agriculture

At the present time about a third of the total land area of the world is being used for agricultural purposes. Another 30 percent is in forests and the remaining land area is in such cover types as desert, tundra, permanently ice-covered land as well as land in such intensive uses as built-up areas including urban places and rural villages and farmsteads, transportation routes and recreational areas.

Land used for agricultural purposes is generally divided into two broad categories. Arable land or cropland is the most important of the two categories since a major part of the food and fiber is produced on this kind of land.[2] Permanent meadows and pastures or grazing land is the other main category of agricultural land. Arable land or cropland is land which is used primarily for the production of crops, although this does not mean that all of the land so classified will necessarily have a crop produced on it each year. Of the 3.6 billion acres presently used as arable land, only about 2.5 billion acres is actually planted to crops in a given year. In subhumid and semiarid parts of the world, extensive areas of land is fallowed in order that sufficient moisture may be accumulated to grow a crop of wheat or barley or some other grain every other or every third year. In humid areas some cropland is left idle each year for various reasons. Also in some humid areas a common practice is to rotate crops with high quality pasture grasses and legumes on which livestock are grazed. Such a rotation has often been used on farms specializing in producing dairy products such as milk and butter.

There is of course a great range in the quality of arable land or cropland used. Some cropland can be used very intensively. In some subtropical and tropical areas of the world where there is freedom or nearly complete freedom from frost, crops may be planted in succession on the same land during the year. In the high latitudes the frost-free season is so short that only a few crops are capable of being matured during the summer. Differences in relief or topography and soil characteristics also make for substantial differences in the quality of cropland from one place to another.

Land used for grazing activities also varies greatly in quality. At one extreme is land in semiarid and arid areas which is grazed by animals only at times when a little forage is available. As many as 50 or more acres may be needed to furnish enough forage for one cow. Extensive areas in parts of the western United States, parts of Australia, the U.S.S.R., and the countries of southwestern Asia have low carrying capacities. At the other end of the quality scale, land used for grazing of livestock in some parts of New Zealand, Australia, north-

---

[2]The term *arable land* has also a connotation of being capable of tillage as well as being land that is presently tilled. Generally the term has been widely used except in the United States as applying to land that is presently being tilled. In the United States, the term *cropland* is much more widely used to designate land used for raising crops.

western Europe, north-central and northeastern United States, Argentina, South Africa, and elsewhere is capable of supporting a cow to each acre of grazing land. Those parts of the world where grazing is possible throughout the year have many advantages in maintaining a high level of productivity. There are also problems of disease and insect infestations not so commonly found in areas with cold winters.

The distribution of the world's cropland was presented in Figure 1.2 and the regional components are shown in Figure 2.1. In looking at the map showing the distribution of cropland, the first impression is the marked high degree of concentration in some places and the complete or nearly complete absence of cropland in other parts of the world. The most extensive areas with little or no cropland are the arid and semiarid areas of Saharan Africa and southwestern Asia, Central Asia, interior Australia, southern Argentina, and western United States. Also the vast cold areas of northern Canada, Alaska, and the U.S.S.R. stand out. (They are unfortunately exaggerated in extent because of the map projection used by the U.S. Department of Agriculture.) The wet tropics of the Amazon Basin of South America and extensive wet tropical areas of central Africa also have relatively little cropland.

On the other hand, the mid-latitude areas of eastern United States and Canada, Europe and the U.S.S.R., eastern China, southeastern Brazil and northeastern Argentina, South Africa, and the southeastern and southwestern parts of Australia have major concentrations of cropland. In the tropics and subtropics, India and southeastern Asia, northeastern Brazil, parts of central Africa, the highlands of Mexico, Cuba, and some other parts of the Caribbean have concentrations of cropland, some of which are likely to be expanded further in the future.

If the cropland area of the world is also examined regionally, it may be noted by looking at Figure 2.1 that Asia (excluding the U.S.S.R.) has the largest area of cropland followed by three other regions which have nearly identical acreages of cropland. These are Africa, U.S.S.R., and Anglo America. Europe, Latin America, and Oceania have the smallest areas. Europe has about a third of the land area in cropland and Asia about a sixth. Australia and Latin America have only four and five percent respectively.

If the cropland area is analyzed as a proportion of total land area from country to country, a great range in the relative importance of cropland may be noted. In TABLE 2.1, large and medium-sized countries and groups of small countries are shown. In a general way this table gives some indication of the places where additional expansion of cropland is likely to be very small simply because so much effort has already been exerted to use all available and worthwhile land for agricultural purposes. On the other hand, those countries using a relatively small proportion of the total land area as cropland should not necessarily be considered as potential areas for a considerable expansion of crop-land. For example, Brazil and some of the other South American coun-tries have more potential than Canada, in the Americas. In Africa the southern part of the continent probably has more immediate potential than the northern Saharan part.

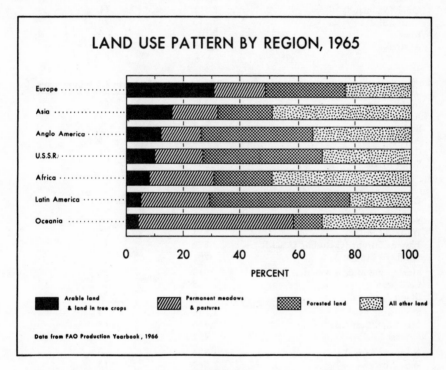

FIGURE 2.1.

If the cropland base of the World of Plenty is compared with that of the World of Poverty, sharp contrast may also be noted. Of the total cropland area of 3.6 billion acres, the World of Plenty had approxi-mately 1.6 billion acres or 45 percent of the total in 1965. Yet the World of Plenty has only 29 percent of the world's population compared to 71 percent living in the World of Poverty, which has to subsist on 55 percent of the world's cropland. Thus the character of agricultural activity is significantly influenced by this fundamental difference in the man-land ratio. There are very striking differences in land use patterns within the two worlds as well as between them.

TABLE 2.1

Cropland as a Proportion of Total Area for Selected
Countries and Groups of Countries, 1965

Countries having an area of more than 1.5 billion acres

	Total Area in Million Acres	Percent of Total Area in Cropland
U.S.S.R.	5.533	10
Canada	2,464	4
China (Mainland)	2,411	11
United States	2,313	20
Brazil	2,102	4
Australia	1,899	5

Countries having an area of more than 0.5 billion acres but less than 1.5 billion acres

India	807	50
Argentina	686	7
Sudan	619	3
Algeria	588	3
Congo (Republic of)	579	21
Saudi Arabia	567	a

Groups of countries having an area less than 0.5 billion acres per country

Scandinavian Europe	285	8
Western Europe	321	32
Eastern Europe (excluding U.S.S.R.)	308	44
Southern Europe	261	42
Mexico and Middle America	619	12
Caribbean	58	18
South America[b]	1,615	4
Southwest Asia[c]	1,145	14
South and East Asia[d]	1,410	18
Central Asia[e]	427	3
Northern Africa[f]	1,834	6
West Africa[g]	1,084	13
East Africa	913	9
Southern Africa	1,619	4
Oceania (excluding Australia)	206	2

[a]Less than 0.5 percent
[b]Excluding Brazil and Argentina
[c]Excluding Saudi Arabia
[d]Excluding Mainland China and India
[e]Includes Bhutan, Mongolia, Nepal, and Sikkim
[f]Excluding Algeria and Sudan
[g]Excluding Republic of Congo

Source: *Production Yearbook, 1966,* Food and Agriculture Organization of United Nations, Rome, 1967, pp. 3-8.

Turning to the change in the area of cropland that has occurred, it is significant to note that expansion of the cropland area has not occurred evenly throughout the world. As a matter of fact, the cropland acreage in the United States has actually declined during this span of time. On the other hand, the U.S.S.R. has made a concerted effort to expand its cropland base into the dry steppes of Siberia. This effort was not very successful because of the unreliable rainfall regime in that part of the country. The United States also had previously expanded its crop production into the western Great Plains in the period from 1915 to 1930 where precipitation conditions are also quite precarious. Some of the decline in cropland area in recent years has occurred along this dry margin of crop production where over-expansion had previously occurred.

Elsewhere in parts of Latin America and Africa and in Australia and New Zealand, expansion of the cropland base has been taking place. Generally this expansion has been on land that is not as good as the land presently used for producing crops, but this is not true in all situations. In the United States, for example, the increase in cropland area in the lower alluvial valley of the Mississippi has been significant after effective flood control had been established. Now many thousand additional acres of crops are being produced on some of the most productive land available in this country.

When the history of agricultural expansion in the United States is reviewed, it becomes obvious that fairly extensive areas of poorly drained land and lands subject to flooding were not developed fully nor effectively during the westward movement of population. In recent years some of these areas such as those of the southeastern coastal plain and the lower alluvial valley of the Mississippi are now being reevaluated and development for agriculture in some places is occurring. The use of new land-clearing and earth-moving equipment has made it possible to develop land which previously could not be improved with more limited equipment.

The first peak acreage of 480 million acres of cropland was reached in 1919 in the United States following the rapid expansion of the cropland base during and immediately after World War I. During the 1930's the first substantial decline in the cropland base occurred. Then another peak was reached in 1949 following World War II, when the cropland acreage climbed to 478 million acres. There has since been a substantial decline in cropland and in 1964 the cropland area was only 434 million acres. This substantial decline in the cropland area has occurred during a period when the total agricultural production has been raised to new heights and when agricultural surpluses have been a major problem for the nation. The Federal Government has instituted several programs to keep surpluses in check and to retire land least suited for growing crops from the cropland base. These programs have only been partially successful. Overproduction of some crops is still taking place and several million acres of land ill-suited to the production of crops are still being used for that purpose.

Equally significant with the leveling off and subsequent decline of the total cropland area of the United States has been the shifting of cropland from one type of land to another. Areas such as the southern Piedmont which formerly had a significant area in crops now has relatively little land used for this purpose. Generally, the cropland area has declined in areas of rough topography where fields are small and ill-suited to the use of large tractor-drawn equipment and where hazards of soil loss are great when row crops are grown. At the same time, level and often poorly drained areas have become significantly more attractive, since it is now possible to remove excess water and control flooding more effectively.

Potential Use of Land for Agriculture

In 1965 the cropland area of the world was estimated by FAO to be 3.6 billion acres, which comprised about eleven percent of the land area. If the basic question about the need for more food production is raised in relation to the amount and distribution of land presently used for producing crops, then the answer to another question must also be sought. How much of the world's land area can be used as cropland in the foreseeable future? Again from a physical standpoint, expansion could extend considerably beyond the limits of reasonable costs of bringing new land into the cropland base in relation to the returns that might be expected from extending the cropland area.

From time to time, qualified experts have attemped to answer this question after making several assumptions about such matters as the level of prices that might be expected from the sale of agricultural products, kind of available technology, and returns to capital and labor invested in developing new cropland compared with returns that might be expected from using existing cropland more intensively.

In recent years more detailed data about soil quality have become available through the World Soil Geography Project in the U.S. Department of Agriculture. These data have permitted one expert, Charles E. Kellogg, Deputy Administrator for Soil Survey of the U.S. Soil Conservation Service, to reexamine an estimate of potential world cropland which he had made after World War II. Dr. Kellogg has concluded that his earlier estimate was too conservative, although at that time many persons thought his estimate was too high. In 1964, Dr. Kellogg

estimated that some 6.6 billion acres were potentially usable as cropland. This is three billion acres more than the acreage being used in
1965, which means that it is possible to think of nearly doubling the
cropland area of the world. Dr. Kellogg concludes: "For a long time
at least, basic soil resources need not be the factor that limits production if soil management is reasonably good."[3]

When the distribution of the present cropland area is compared with
the area potentially usable as cropland, the most striking large area
suitable for cropland use is in the forested humid tropics and subtropics, particularly in Africa and South America, where there are
excellent soils available for crop use. Some extensive forested areas in
the mid-latitudes, particularly the United States for example, are also
potentially usable for crop production. Eastern and southern Asia has
relatively little unused arable land. In some parts of the world, particularly in the United States, Australia, New Zealand, and parts of
Latin America, there are extensive areas of land presently used for
grazing which could be used for crop production. Since some food is
presently being produced from this grazing land, the gain in additional
food supplies to meet present needs will not be as great as from new
cropland developed in forested areas, from which very little food is
presently obtained.

In contrast to the areas just discussed as having more land that
could be used as cropland, it must be remembered that approximately
one-half of the land area of the world is not suited for crop productions.
Nor is there much prospect that this land will be useful for producing
crops in the foreseeable future. The extensive areas of permanent ice
and snow, the large expanses of tundra, the high and rugged mountain
areas, and the arid and semiarid areas of the world that lack water
for irrigation are the lands of little hope for agriculture.

When the distribution of the world's population is compared with
the distribution of land that is potentially usable as cropland, a striking question may appropriately be raised. How can the extensive areas
of potential cropland located in areas of relatively sparse population
be used effectively in producing food that is needed elsewhere in the
World of Poverty. Movement of food in international trade is apparently not the answer, since the countries needing surplus food produced in other parts of the world do not generally have the foreign
exchange for the purchase of food elsewhere. Migration of people from
areas of deficit food supplies across international boundaries to countries having potential cropland available for development has also been

---

[3]Charles E. Kellogg, "Potentials for Food Production," *Farmer's World: The
Yearbook of Agriculture, 1964.* Washington, D. C., U.S. Department of Agriculture,
p. 61.

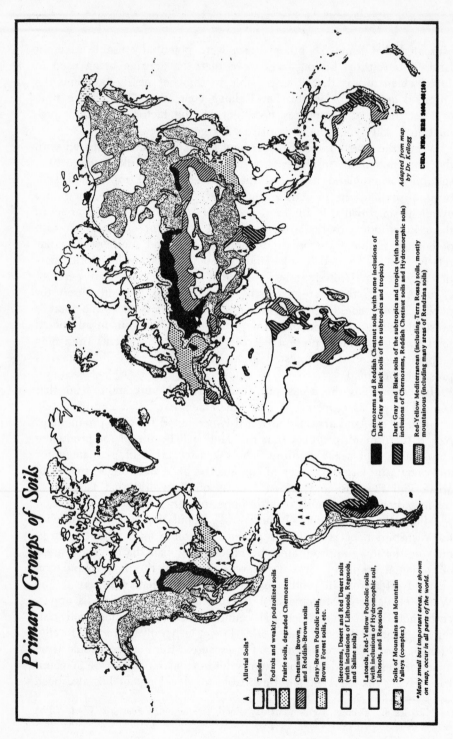

Primary Groups of Soils

Ice cap

▲ Alluvial Soils*

Tundra

Podzols and weakly podzolized soils

Prairie soils, degraded Chernozem

Chestnut, Brown, and Reddish-Brown soils

Gray-Brown Podzolic soils, Brown Forest soils, etc.

Sierozems, Desert and Red Desert soils (with inclusions of Lithosols, Regosols, and Saline soils)

Latosols, Red-Yellow Podzolic soils (with inclusions of Hydromorphic soil, Lithosols, and Regosols)

Soils of Mountains and Mountain Valleys (complex)

*Many small but important areas, not shown on map, occur in all parts of the world.

Chernozems and Reddish Chestnut soils (with some inclusions of Dark Gray and Black soils of the subtropics and tropics)

Dark Gray and Black soils of the subtropics and tropics (with some inclusions of Chernozems, Reddish Chestnut soils and Hydromorphic soils)

Red-Yellow Mediterranean (including Terra Rossa) soils, mostly mountainous (including many areas of Rendzina soils)

Adapted from map
by Dr. Kellogg

USDA NEG. ERS 2699-64(10)

FIGURE 2.2.

Source: A Graphic Summary of World Agriculture, p. 6, U.S. Department of Agriculture, 1964.

28

TABLE 2.2
Estimated Potentially Arable Land in the World

Soil Group on Accompanying Soil Map	Percent of Soil Group Potentially Arable	Acres Potentially Arable (millions)
Prairie Soils, Degraded Chernozems	80.0	242
Chernozems and Reddish Chestnut	70.0	660
Dark Gray and Black Soils of Subtropics and Tropics	50.0	618
Chestnut, Brown, Reddish Brown	30.0	892
Sierozems, Desert	.5	34
Podzols and Weakly Podzolized	10.0	320
Gray-Brown Podzolic	65.0	972
Latosols, Red-Yellow Podzolics	35.0	2,780
Red-Yellow Mediterranean	15.0	41
Soils of Mountains	.5	30
Tundra	.0	0
Total	---	6,589

These criteria were used to define arable land:

That reasonably good management would be used including appropriate combinations of adapted crop varieties, water control methods, pest control, and methods of plant nutrient maintenance, including some chemical fertilizers.

Crops include the ordinary food, fiber, and industrial crops that are normally cultivated as well as fruits, nut crops, rubber, sisal, coffee, tea, cocoa, palms, vines, and meadow crops that may or may not be cultivated.

All regular fallow land is counted, including the natural fallow under shifting cultivation.

Irrigation of arid soils is limited by water from streams and wells. Sea water is excluded as a potential source.

Source: *Farmer's World: The Yearbook of Agriculture,* 1964, Washington, D. C.: U.S. Department of Agriculture, p. 62.

a dead-end road in recent years. Several countries such as the United States and Australia, for example, have had strict immigration laws which effectively deter such migrations. Even if greater movement of population from the World of Poverty to the World of Plenty were possible, a significant improvement in the man-land ratio in the country from which people emigrated might not necessarily occur.

Turning to the United States for a more detailed look at the possibilities for expanding the cropland area, it is readily apparent that this country still has a considerable amount of land which is potentially capable of being used as cropland. Much of this land is presently a part of the farmland area of this country. It is land in farms that is presently being used for grazing or is forested. Very little of the potential cropland is still owned by the Federal Government. Although about a third of the total land area of this country is still Federally owned, very little of this land has much value for cropland use. Some extensive areas of forest land now being managed on a sustained yield basis and which are producing valuable forest products could be used as cropland if needed.

A National Inventory of Soil and Water Conservation Needs completed in 1959 for the United States gives a fairly reliable picture of the capabilities of the nation's land in relation to the uses being made of land resources at that time. TABLE 2.3 relates land use to land capability. In making the inventory, land was classified into eight categories with Class I land being the best land for agricultural uses and Class VIII land being nearly worthless for such use. Land placed in Classes I, II, and III is suitable for regular cultivation and other uses. Class IV land is suitable for occasional cultivation and other uses. Land placed in Classes V to VIII is generally not suitable for cultivation but is suitable for other uses such as grazing, forestry, and recreation.

TABLE 2.3

Land Capability and Land Use, 1959*

Capability Class	Cropland	Grazing Land	Forest and Woodland	Other Land	Total
	Million Acres	Million Acres	Million Acres	Million Acres	Million Acres
I	27.4	3.9	3.6	1.2	36.1
II	192.8	42.8	43.2	11.3	290.1
III	152.9	66.5	77.6	13.8	310.8
IV	48.9	53.9	58.1	7.8	168.7
V	1.8	10.5	28.9	1.8	43.0
VI	17.9	166.1	87.9	4.9	276.8
VII	5.6	138.4	142.7	7.5	294.2
VIII	0.1	2.5	6.4	17.7	26.7
Total	447.4	484.6	448.4	66.0	1,446.4

*For all land except that owned by the Federal Government and land in urban and built-up areas.

Source: *Agricultural Land Resources: Capabilities, Uses, and Conservation Needs,* Agriculture Information Bulletin 263, U.S. Department of Agriculture, Washington, D. C., 1962.

The main message gained from making a comparison of land use with land capability is the assurance that land resources in this country are quite adequate to accommodate a significant expansion of the cropland area. There are 113 million acres of grazing land and 124 million acres of forest and woodland in land capability Classes I, II, and III which is capable of being used as cropland. Some but not all of the 25 million acres of other land in capability Classes I, II, and III could also be used as cropland. Some of this land is occupied by farmsteads and other uses not easily converted to cropland. Land in capability Class IV presently being used as cropland probably represents a maximum that should be so used at any one time in that capability class. It must also be noted that about 25 million acres of cropland is still situated on land in capability Classes V, VI, VII, and VIII. Since none of this land should really be used as cropland from the standpoint of maintaining effective conservation control, there is only a net balance of about 212 million acres presently used as grazing land and forest and woodland that could be used as cropland if needed.

However, as previously noted, the acreage of cropland has been consistently declining since the early 1950's and will possibly continue to do so for at least a few more years. This decline in the cropland base has to some extent been the result of the rapid expansion of urban and transportation uses but the expansion of such uses of land does not account for nearly all of the decline that has occurred in recent years. A more significant reason why the decline has continued is the rapid substitution of capital for land and labor which is still continuing in American agriculture. Investing capital in more intensive use of the present cropland is generally more profitable than investing capital in clearing, draining, irrigating, and otherwise improving land not presently a part of the cropland area.

Technology and the Agricultural Use of American Resources

The frontier of American agriculture over the past half century has been a technological frontier. Practically no net expansion of the cropland base has occurred and there has even been a contraction of the cropland area. This has been partly brought about by Federal subsidy, used to encourage farmers to put their land in the soil bank and other temporary land retirement programs. Some significant shifts in the type of land used for the production of crops has also been taking place as previously noted. The very fact that agricultural production has increased so substantially in the United States since 1940 on a contracting cropland base is a striking demonstration of the significant role being played by technological innovations in American agriculture.

One of the most compelling forces which has led to the ready acceptance of technological advances by the American farmer has been the rising cost of labor which can find better wages in manufacturing, service, and other activities. In order to stay in business, the American farmer has had to accept labor-saving devices as well as many technological innovations which enable him to get better yields from his crops and a better return from his livestock. The accelerating use of technological innovations has brought about a substantial increase in

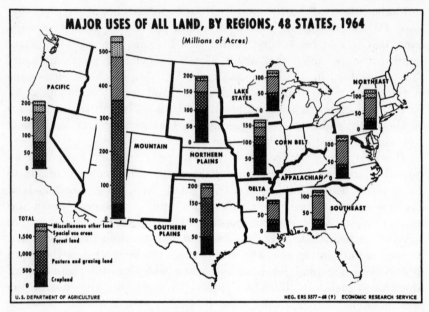

FIGURE 2.3.

Source: *Major Uses of Land and Water in the United States with Special Reference to Agriculture: Summary for 1964.* p. 10, U.S. Department of Agriculture, 1968.

the size of farm and in the kind of operation which the farmer must have in order to make a profit in farming. In 1900, the average size of farm in the United States was 147 acres. By 1964, it had more than doubled and was 352. The capital investment in machinery, fertilizers, insecticides, pesticides, better seed, and other technical improvements has also increased very substantially. Similar changes in general characterize the business of agriculture in other countries of the World of Plenty.

In sharp contrast, the World of Poverty is still characterized for the most part by the intensive use of labor on limited land resources. One of the most striking differences between the agriculture of the World of Plenty and that of the World of Poverty is that in the former intensive use of capital is much more characteristic and in the latter intensive use of labor predominates in the agricultural sector of the economy. Very little capital is available to use for the purchase of modern technology which might bring about significant increases in agricultural output if instituted on farms in the World of Poverty. It is not likely that capital invested in the same improvements and innovations currently characterizing agriculture in the World of Plenty would bring the same returns. Money invested in better seeds rather than in a labor-saving tractor might be a better initial use of limited capital in a country where labor is plentiful and for which there is little immediate alternative employment.

What are some of the ways in which the ready acceptance of technological advances have brought about a substantial enhancement in the productivity of the American farmer and of American farmland? The technological revolution in agriculture has been dramatic during the past 25 years and changes have been taking place at an accelerated pace. However, the foundation of experimentation and education which has enabled this country to make such rapid progress in agriculture in recent years was established much earlier. The passage of the Morrill Act in 1862 set the stage for State and Federal cooperation in financing experimentation and education. The agricultural experiment stations and agricultural extension service have played major roles in agricultural research and education respectively. American corporations engaged in the manufacture of machinery, fertilizers, pesticides and insecticides, and other materials needed in agriculture have also made major contributions to the success story of American agriculture. Some specific examples of the impact of technology on agriculture are appropriately cited here.

1. *Substitution of inanimate for animate power and introduction of auxiliary equipment.* The introduction of the tractor has undoubtedly been one of the greatest single advances in American agriculture over the past 50 years. It has been estimated that this substitution of the tractor for the horse and mule on American farms has freed approximately 80 million acres of cropland for the production of food for human consumption, since this cropland is no longer needed for feeding a horse and mule population which in 1919 (the peak year) totaled 21 million. This change in the source of power for American farming has involved really the substitution of one basic resource for another—oil for soil. Petroleum reserves are now being used more rapidly because of this change, which has brought such an improvement in the efficiency of labor employed in agriculture in the United States.

In the 1820's, before the invention of the McCormick reaper in 1834, about 50 to 60 man-hours of labor were required to produce one acre (with a 20-bushel yield) of wheat with walking plow, use of brush for harrowing, hand broadcasting of seed, and harvesting with scythe and flail. By the 1890's, this labor requirement was reduced to eight to ten man-hours of labor to produce one acre (20-bushel yield) of wheat with gang plow, seeder, harrow, binder, thresher, wagons, and horses. By the 1920's, this requirement was reduced further to only three to four man-hours of labor to produce one acre (20-bushel yield) of wheat with the use of three-bottom gang plow, tractor, 10-foot tandem disk, harrow, 12-foot combine, and trucks. Today, wheat is being produced with one to three man-hours of labor (1 acre with 25-bushel yield) with the use of large tractors, 10-foot one-way disk, 12-foot rod weeder, harrow, 14-foot drill, 20-foot self-propelled combine, and trucks.

2. *Greater use of fertilizers.* Fertilizers were first used extensively in the American South in the production of cotton and tobacco. Prior to the 1940's, relatively little fertilizer was used in the Midwest. Today fertilizers are widely used in increasing amounts in producing corn in the United States. Relatively little fertilizer is being used in producing wheat in the Great Plains because of the risk of inadequate moisture. The average annual consumption of commercial fertilizers for the period 1935-39 for the United States was seven million tons. The average annual consumption for the 1950's was 22 million tons, and for the period 1960 to 1964 the annual average consumption had risen to 27 million tons. This four-fold increase in the use of fertilizers has contributed substantially to higher yields for practically all crops.

3. *Introduction of better varieties of plants and better breeds of animals.* In many ways the development of new and better varieties of plants and better breeds of animals has contributed to the quality and quantity of agricultural products produced. American farmers now produce at least 20 percent more corn from 25 percent fewer acres than were used for corn production in 1930 before the widespread use of hybrid corn. In addition to increased yields attributable directly to hybridization, the hybrid varieties are also more efficient users of fertilizers. Some hybrids have been developed which are more resistant to insects and diseases. Also very important is the better adaptability of hybrid varieties to mechanization because of a more uniform maturing time and greater resistance to lodging (falling to the ground). Planned breeding with scientific methods has contributed greatly to the development of new varieties of wheat which yield better, utilize fertilizers more efficiently, and have greater resistance to rust and other diseases.

For some crops the search for new varieties has been directed specifically toward obtaining a variety which can be mechanically harvested. Particularly in the production of fruits and vegetables, the problem of obtaining labor at the critical harvest time has been an extremely difficult one to resolve. The tomato is an excellent example of the impact of mechanical harvest upon the production of an agricultural commodity. Farmers were actually abandoning the production of tomatoes for processing in California. Then the invention of a mechanical tomato harvester was joined by the development of a tomato variety that had small vines, a concentrated setting and ripening period, and a pliable fruit which would resist bruising when harvested mechanically. In 1966, the mechanical tomato harvester eliminated approximately 3.6 million man-hours of labor.

Many examples of higher productivity in livestock are also easily cited. Because of the relatively high per capita consumption of livestock products in the United States, developments in livestock breeding have been especially significant. In the breeding of hogs, one of the major objectives has been to develop a hog with more lean meat and a minimum of fat, since vegetable oils have now largely replaced lard in cooking. In the case of sheep, the breeding efforts were directed toward getting a breed that would produce a wool of fine quality and also at the same time a meat of good quality. By crossing fine-wool breeds such as the Merino and Rambouillet with the Lincoln, a new breed, known as the Columbia, was developed for the western sheep industry.

In trying to get a satisfactory breed of beef cattle for the American South, Zebu-type cattle were imported from India. These cattle can stand heat better because they have sweat glands and are more tolerant of insects because they can twitch their skin. However, the quality of their meat was not as good as that obtained from beef breeds introduced into this country much earlier from the British Isles. Furthermore, they were not as reliable in reproduction, which is an important matter in producing beef profitably. Several attempts at developing crossbreeds between the Zebu and British breeds were made. One of the first attempts was made on the King Ranch in Texas where the crossing of the American Brahman (developed from several Indian Zebu breeds) with the Shorthorn resulted in the development of the Santa Gertrudis, which is now widely raised in the United States and in other countries.

4. *Better control of insects, diseases, and weeds.* Many problems have confronted the farmer in keeping insects and diseases under control. The boll weevil nearly wrecked the southern economy in the short span of a few years after first being introduced into Texas from Mexico in 1892. For awhile it was the world's largest consumer of raw cotton. Now by spraying and dusting of poisons (in many areas from low-flying crop duster planes), the boll weevil has been brought under control. The cattle tick, the corn borer, the Japanese beetle, viruses, wheat rust, nematodes, Mediterranean fruit fly, and a host of other insects and diseases have plagued farmers in different parts of the country from time to time.

Weeds are also a major problem in farming. Although the dictionary defines a weed as a plant of no value, such a definition needs some qualification. Johnson grass, which has long been despised by the cotton farmer as a major problem, is considered by beef cattle producers in the South to be one of the most efficient grasses for feeding cattle. Kudzu, a legume introduced into the South from the Far East to help control gully erosion along highways and in badly eroded fields, has now been labeled an undesirable plant by foresters, because it climbs into tree tops and chokes out the tree foliage and eventually results in the tree dying.

Effective weed control has been developed for some weeds. For others, such as the water hyacinth in Florida, the search still goes forward. Over the years farmers have used many approaches to control weeds such as hoeing, hand-pulling, mowing, burning, planting smother crops, and rotating crops. Other attacks have been made in order to keep weeds under control. Laws have been passed to penalize farmers who continue to permit certain types of weeds to grow and multiply. Legislation has also been passed to prohibit the sale of crop seeds which have noxious weed seeds intermixed with them. Inorganic and organic chemicals have been used. One of the highly effective chemicals used widely in recent years is 2, 4-D, which is an organic chemical capable of killing susceptible plants in a highly efficient manner. This herbicide was first introduced in 1944 and since then there has been a remarkable increase in the use of selective herbicides in agriculture.

5. *Better tillage and conservation practices.* With the mechanization of farming it became possible to improve the tilling of crops by making tillage operations more timely and effective. Plowing land or tilling crops with horses and mules took much more time and, often, the crops were not cultivated before weeds were able to get a head start or while moisture conditions were favorable. Now with tractor power these operations can be completed quickly and efficiently.

Much progress has also been made in recent years in managing land resources in such a way that these resources are not wantonly wasted and so the land will continue to yield agricultural products efficiently on a sustained basis indefinitely. During the early settlement period in America when land was abundant, farmers often used fields until yields started to decline and then they would clear other land and even start over again on a new farm. Today, forms of shifting cultivation are still in use in some parts of the world but in America such an approach to farming has long since ceased to exist for the most part.

Now farmers are generally quite receptive to land-use practices which will enable them to conserve their soil resources and at the same time enable them to produce agricultural commodities efficiently in competition with other farmers. Planting crops on the contour in alternating strips of close-sown and cultivated row crops with the strips placed at right angles to the direction of slope has enabled farmers to use much gently to moderately sloping land that could not otherwise be used. Installation of proper drainage to get excess water off valuable cropland quickly, use of grass-planted waterways to permit water to flow into natural drainage channels, rotation of crops particularly on certain types of land, stubble mulching in the Great Plains where wind erosion is a problem, and many other practices are now being widely used to conserve soil and water resources used in agriculture.

Needed: An Agricultural Revolution for the World of Poverty

Over the span of many years, even hundreds of years, the production of food crops and livestock has for the most part developed along traditional lines in countries which comprise the World of Poverty. Over the centuries, peasant farmers have continued to use the same crops, the same varieties of seed, and the same farming practices as those used by their fathers and grandfathers and by their ancestors of a long time ago. The practices and crop varieties that have sustained life down through the years have proved to be the most reliable approach for those who lack the education to understand how other practices and other varieties might yield better results. When life is being sustained close to a level of bare existence, experimentation with new ideas may prove fatal. A willingness to accept technological innovation develops slowly and only after new techniques have clearly been demonstrated to be effective.

In recent years new varieties of rice, wheat, and corn have been developed. Since each of these crops is a major staple in different parts of the World of Poverty, the introduction of new high-yielding varieties of these crops could make a major impact upon the agriculture of the developing world if they are readily and widely used. Many of the people of developing countries are keenly aware of their plight. If clearly demonstrated results can be observed by them and if the peasant farmer can obtain the necessary money to acquire the new seed, fertilizers, and pesticides, he will in many instances be willing to break with tradition. As recently as 1964-65, only 200 acres of the new rice and wheat varieties developed in the Philippines and in Mexico respectively were planted in Asia. By 1967-68 these varieties were planted on about 20 million acres according to estimates made by the International Agricultural Development Service of the U.S. Department of Agriculture.

This is indeed a remarkable initial impact and the prospect that countries such as the Philippines and Pakistan may soon become net food exporters instead of net food importers is a most impressive step toward improving levels of living in parts of the World of Poverty.

However, it will not be an easy matter to sustain the momentum of this striking initial impact when the diverse character of the established traditional agriculture is surveyed. New problems will inevitably arise with the introduction of new technology and some of the existing constraints will prove troublesome. Among these constraints is the limited area of irrigated land that is available and which will need to be expanded if the use of new varieties of rice in particular is to be effective. Lack of capital will of course be an impediment to the construction of new irrigation facilities, since there are other pressing demands upon the scarce capital resources available.

There is also reason for considerable concern about the existing market structure in many areas and the ability of these markets to absorb significant increases in output at prices which will be conducive to continuing the accelerated rate of grain production. Purchasing power of the urban masses in most developing countries is extremely limited and, unless labor productivity in the non-agricultural sector of the economy can be increased, the gains in the agricultural sector may be short-lived. Improved and enlarged storage facilities and better transportation networks are also likely to be vital prerequisites to the establishment of more effective marketing arrangements for farm products in developing nations.

Unless concurrent investments are made in education and health as well as in the development of technical skills which will provide a sound basis for enhancing the productivity of the population, the impact of new agricultural techniques may not bring lasting improvements in the level of living. With limited capital available, it is often not easy to make such investments in the intangibles of a higher level of education and improved health conditions.

A vexing problem for some parts of the World of Poverty that must be resolved before a technological revolution in agriculture will be effective is the institutional problem of land reform. In many parts of Latin America, major changes are needed in the distribution of available land resources among the landless peasants if adequate economic incentives are to be provided for raising the level of agricultural production.

Another vital link in the complex set of conditions required to bring about an agricultural revolution in the World of Poverty is the establishment of an effective system of credit which will permit farmers to obtain both short-term loans for production expenses and longer term loans for capital improvements. This credit will need to be available at interest rates which are reasonable and which farmers will be able to afford.

Perhaps the most critical issue of all the many complex problems facing the developing world in bringing off an agricultural revolution will be the ability to maintain the momentum of technological innovation and to diversify and deploy the use of technology so that its benefits will permeate to the very roots of the poverty of the peasant farmers, which needs to be overcome. Qualified scientists and trained technicians will be needed in increasing numbers in order that the benefits of science and technology may be extended to all parts of the developing nations which are striving to attain levels of agricultural and industrial production that will permit them to become a part of the World of Plenty.

REFERENCES

A Chronology of American Agriculture, 1790-1965, Washington, D. C.: U.S. Department of Agriculture, 1965.

Agricultural Land Resources: Capabilities, Uses, and Conservation Needs, Agriculture Information Bulletin 263, Washington, D. C.: U.S. Department of Agriculture, 1962.

Farmer's World: The Yearbook of Agriculture, 1964, Washington, D. C.: U.S. Department of Agriculture.

Power to Produce: The Yearbook of Agriculture, 1960, Washington, D. C.: U.S. Department of Agriculture.

Science for Better Living: The Yearbook of Agriculture, 1968, Washington, D. C.: U.S. Department of Agriculture.

Chapter 3

Patterns of Crop and Livestock Production

Agriculture as an economic activity is concerned with producing crops and raising livestock. Crops and livestock are divided into two main classes, those used as food and those not used for food. These classes in turn consist of several subclasses which are listed here along with the main crops and livestock or livestock products in each subclass:

Food Crops

Cereals—wheat, rye, barley, oats, corn (maize), millets, sorghum, and rice

Sugar—sugar cane and sugar beets

Starchy roots—potatoes, sweet potatoes, yams, and cassava

Pulses—dry beans, dry peas, dry broad beans, chickpeas, lentils, pigeon peas, cowpeas, vetch, and lupins

Oil crops—olive oil, palm kernels, palm oil, soybeans, peanuts (groundnuts), cottonseed, rapeseed, sesame seed, sunflower seed, and copra (coconut meat)

Vegetables, fresh—tomatoes, cabbages, cauliflowers, green beans, green peas, sweet corn, onions, and many others

Fruits—oranges, grapefruit, lemons, limes, tangerines, apples, pears, plums and prunes, grapes and raisins, peaches, apricots, figs, bananas, pineapples, dates, and many others

Nuts—pecans, English walnuts, almonds, Brazil nuts, black walnuts, and many others

Wine—mainly from grapes but made from some other fruits as well

Cocoa—from cacao

Nonfood Crops

Beverage—coffee and tea

Industrial oilseeds—castor beans, hempseed, and linseed (from flax)

Tobacco
Fibers—cotton lint, flax fiber, hemp fiber, jute, sisal, henequen, abaca, and others
Rubber—(natural)

Livestock for Food

Meat—cattle and calves, swine (hogs), sheep and lambs, poultry and other animals
Eggs—chicken and others
Milk—mainly from cows, goats, sheep, buffalo
Honey

Livestock, Nonfood

Sheep for wool
Goats for mohair
Cocoons for silk

Although a great variety of crops and livestock are used for food, fiber, and other purposes around the world, some crops and livestock are of major importance and are more universally produced than others. As shown in Figure 3.1, the cereal grains alone account for about 71 percent of the world's harvested crop area. Of the grains, wheat, rice, and corn are the most important and are widely recognized as the staple food for millions of people in different parts of the world. Corn is particularly important as a staple food crop in Latin America, and rice is well known for its importance in southern and eastern Asia. Wheat is a major staple in Anglo America, Europe, and the U.S.S.R., but it also is important in parts of Pakistan, India, China, southwestern Asia, northern Africa, Argentina, Australia, and elsewhere as well.

Crop and Livestock Production in the United States

In the United States five crops, corn, wheat, soybeans, sorghums, and oats, accounted for two-thirds of the acreage of crops harvested in 1967. All hay crops accounted for another fifth of the acreage. Cotton, which is no longer nearly so significant a crop in American agriculture as formerly, still ranks high in value. Fruits and vegetables, which occupy a relatively small acreage, are also very important from the standpoint of value.

Within the present century three crops, which were previously of little importance, have emerged as major crops on American farms. These are soybeans, sorghums, and alfalfa. In 1919 only 112,000 acres of soybeans were harvested for beans compared to nearly 40 million acres harvested in 1967. Sorghums have expanded during the same period from five to 18.5 million acres. Alfalfa has increased from two million acres in 1900 to more than 28 million acres in 1967. At the same time other crops have declined significantly in their importance as products of American farms. In 1919, oats occupied nearly 38 million acres of land and were at that time an important feed for horses and mules. In 1967, this crop was harvested from only about 16 million acres. Clover and timothy, also widely used as a hay crop particularly fed to horses and mules, declined from 37 million acres in 1909 to less than 13 million acres at present.

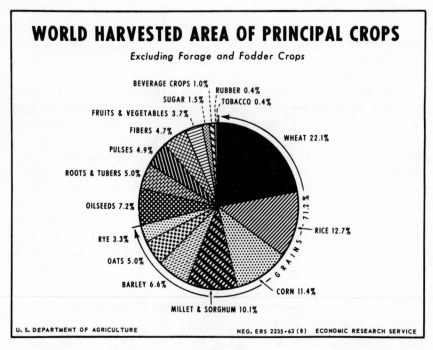

FIGURE 3.1.

Source: L. R. Brown, *Man, Land, and Food: Looking Ahead at World Food Needs*, p. 21, U.S. Department of Agriculture, Washington, D. C., 1963.

Looking at the shifts in crop production which have occurred more recently as shown in TABLE 3.1, one is impressed particularly with the remarkable increase in the acreage of soybeans harvested and with the sharp decline in cotton. Also a substantial drop in corn acreage and an increase in wheat acreage has also occurred. In order to understand these shifts fully, one must be familiar with the various Federal programs and subsidies which are used to control acreage and production

TABLE 3.1
Crops Harvested, 48 States, 1954-67

Item	1954	1959	1964	1967†
	Million Acres			
Food Crops				
Food Grains				
Wheat	54.4	51.7	49.8	59.0
Rice	2.5	1.6	1.8	2.0
Rye	1.8	1.4	1.7	1.1
Total Food Grains	58.7	54.7	53.3	62.0
Irish potatoes	1.4	1.3	1.3	1.5
Sweet potatoes	.3	.3	.2	.1
Dry beans	1.5	1.4	1.4	1.2
Dry peas	.3	.4	.3	.2
Cowpeas for peas	.3	.2	.1	.1
Sugarcane, all	.4	.5	.7	.5
Sugar beets	.9	.9	1.4	1.1
Peanuts for nuts	1.4	1.4	1.4	1.4
Soybeans for beans	17.0	22.6	30.8	39.7
Fruits and planted nuts*	4.5	4.6	4.6	4.8
Commercial vegetables	3.8	3.4	3.3	3.6
Total Food Crops	90.5	91.7	98.8	116.3
Feed Crops				
Feed Grains				
Corn	80.2	81.9	65.4	69.7
Oats	40.6	27.7	19.7	15.9
Barley	13.4	14.9	10.3	9.2
Sorghums, all	18.1	19.0	15.8	18.5
Total Feed Grains	152.3	143.5	111.2	113.3
All Hay				
Tame	59.6	55.6	56.8	55.2
Wild	12.9	10.7	10.5	9.5
Total Hay	72.5	66.3	67.3	64.7
Total Feed Crops	224.8	209.8	178.5	178.0
Other Crops				
Cotton	19.3	15.1	14.1	8.1
Flaxseed	5.7	2.9	2.8	2.0
Tobacco	1.7	1.2	1.1	1.0
Broomcorn	.3	.2	.2	.1
Sweetclover seed	.3	.1	.1	.1
Timothy seed	.3	.3	.2	.2
Minor crops†	2.2	2.0	1.6	1.6
Total Other Crops	29.8	21.8	20.1	13.1
Total Crops Harvested‡	345.1	323.3	297.4	307.4

*Includes tree fruits, small fruits, and planted nuts.

†Consists of small acreages of certain vegetables and field crops not included in the 59 principal crops, various legumes, and other crops harvested by livestock, principal crops in minor producing States, and nursery and greenhouse products.

‡Excludes duplicated acreage in alfalfa, red clover, alsike clover, and lespedeza seeds harvested from hayland, peanut vine hay harvested from land where peanuts were harvested, velvetbeans, and several minor crops.

Source: *Agricultural Statistics* (1954-68): Washington, D. C., U.S. Department of Agriculture.

of crops which have been overproduced in most years since 1954. Soybeans is a major exception and no controls have been placed on the production of this crop. Strong domestic demand coupled with good export markets in western Europe and Japan have meant consistently good soybean prices. On the other hand, cotton has continued to be in chronic over-supply because of competition from synthetic fibers and because of increased production in several other parts of the world. In 1967, wheat acreage controls were relaxed because of serious droughts in India and Australia, which sharply reduced American reserves of wheat. Corn continues to be a remarkable example of increased productivity on fewer acres of land. In 1954, corn harvested for grain from 66.8 million acres totalled 2.6 billion bushels. In 1964 the Bureau of the Census reported 3.4 billion bushels harvested from only 53.8 million acres.

In the United States, livestock and livestock products accounted for 53 percent of the total value of all farm products sold in 1964. Cattle and calves and dairy products sold are in turn responsible for about two-thirds of the value of all livestock and livestock products sold. Poultry and poultry products are now more important than hogs and pigs; and sheep, lambs, and wool sold are next in importance. The total value of livestock and livestock products sold in 1964 was as follows:

	Value of Sales (millions of dollars)
Dairy products	4,637
Cattle and calves sold	8,162
Poultry and poultry products	3,063
Chickens sold	(1,038)
Eggs sold	(1,461)
Turkeys, ducks, and geese and their eggs sold	(564)
Hogs and pigs	2,334
Sheep, lambs, and wool	454
Sheep and lambs	(354)
Wool	(100)
Goats, mohair, horses and mules, and all other	191
Total	18,841

The feed needed to produce these livestock and their products is derived from three main sources: (1) 923 million acres of pasture and grazing land, 200 million acres of which are Federally owned; (2) almost 65 million acres of hay; and (3) about 113 million acres of feed grains. Thus of the 297 million acres of land used for harvested crops in 1964, about 178 million acres was used to produce feed for livestock. The various types of livestock used this feed supply as follows:

Percent

Dairy cattle 26
Cattle other than dairy 39
Hogs 16
Poultry 10
Sheep, goats, and all other 9

About 55 percent of the feed used in producing livestock in 1964 came from pasture, hay, and other forage crops; another 33 percent of feed was comprised of corn, sorghums, oats, barley, and other feed grains; and 12 percent of the feed was derived from milling by-products and the cake of oil seed. The type of feed consumed by the main kinds of livestock differs greatly as shown by the following percentages:

Percent used by

	Dairy Cattle	Hogs	Poultry	Cattle Other than Dairy
Feed grains, milling by-products, and protein concentrates	18	34	22	19
Pasture, hay, and forage	33	1	negligible	57

During the past 20 years some major changes have been taking place in the production of livestock in the United States. There has been a significant decline in the number of farms with livestock and the proportion of the farms that are raising each kind of livestock has decreased sharply. For example in 1940, 76 percent of all farms kept milk cows; but by 1964 only 36 percent of all farms had milk cows. In 1940, 62 percent of all farms had hogs but in 1964 only 34 percent were raising hogs. Likewise, the percent of all farms raising chickens dropped from 85 to 38 percent during the same period.

There are several reasons for this fundamental change in the pattern of livestock production which is resulting in a high degree of concentration of this production on a relatively small proportion of American farms. The rapid introduction of technology has increased significantly the amount of cropland that a farmer can operate but a similar impact of technological innovation has not occurred in livestock production. Another major factor has been the several programs of crop price supports which have made the production of crops for sale less of a risk than producing livestock. Third, the rapid decline in the farm population has left a very limited and unreliable supply of farm labor for the care of livestock. Also the opportunities for farmers or members of their families to obtain off-farm employment have competed with the requirements for labor in producing livestock.

Major World Crop Patterns

Several of the major crop and livestock patterns are shown on the following pages in a series of maps which appeared originally in *A Graphic Summary of World Agriculture* prepared by Nelson P. Guidry of the U.S. Department of Agriculture. A brief analysis of the patterns accompanies each map. This analysis of some of the major agricultural products will be followed in the next chapter by a geographical discussion of systems or types of agricultural production in order that regional combinations of crop and livestock production may also be appreciated.

wheat

Wheat is the world's leading crop both in acreage grown and tons produced. In 1965 the world acreage of wheat totalled 540 million acres and production amounted to 266 million metric tons. It is widely produced in the world's agricultural regions; the main exception is the absence of wheat from areas of wet tropical and subtropical climates where high humidity conditions are conducive to some of the major diseases of wheat. Over the past 15 to 20 years, both the acreage used for wheat and the production of wheat have increased significantly, particularly in the U.S.S.R. For the five-year period of 1948-52 the world annual average acreage of wheat was 420 million acres. For the five-year period 1961-65 the average annual area in wheat was 519 million acres, an areal increase of about 24 percent. Production increased even more spectacularly with a change during the same time period from 171 million metric tons to 253 million metric tons, an increase of nearly 50 percent. Europe and the U.S.S.R. have had the most substantial increases in production, the former because of significant increases in yields and the latter more because of substantial increases in area. However, all parts of the world where wheat is produced have shared in the production increases.

Bread has long been a staple in the diet of millions of people living in the middle latitudes including areas having a dry summer subtropical (Mediterranean) type of climate. Since the best bread can be made from wheat, this crop has had wide acceptance as a basic source of food. Wheat was probably first grown in parts of southwestern Asia but its culture has spread rapidly to many parts of the world.

Climatically, wheat must have a growing season of at least 100 days and probably the most desirable amount of annual precipitation for its culture is about 32 inches. However, in the Big Bend area of the State of Washington and in Australia, wheat is produced under an annual precipitation regime of ten inches. It is also extensively produced in the northern Great Plains of the United States and Canada as well as in Argentina and South Africa where annual precipitation is only about 15 inches. In the southern Great Plains of the United States where evapotranspiration rates are high, about 17 inches of rain are needed annually to produce wheat economically. The economies of scale permit areas, such as the Great Plains, to produce wheat in a climatic regime not ideally suited for that crop but better suited for the production of small grains than a crop such as corn. Large farms with fully mechanized operations result in great savings in labor that make wheat the most profitable crop in the Great Plains of the United States and Canada, Australia, Argentina, and the U.S.S.R. where the population is relatively sparse.

Two main approaches are used in the production of wheat depending on the severity of winters. Winter wheat is planted in the autumn and is harvested in early summer in areas having sufficiently mild winters such as the southern Great Plains and southern and western Europe. Winter wheat begins its growth before freezing weather and some farmers with livestock will use their wheat fields for some fall and winter grazing. Such grazing does not damage the wheat if the animals are taken off when growth restarts in spring.

Spring wheat is planted in spring and is harvested in early autumn or late summer. It is grown where winter temperatures are too severe for wheat to be planted in the preceding autumn. Generally, yields of spring wheat are not as good as for winter wheat. Thus, wherever possible, winter wheat is grown.

Cultivated summer fallowing or dry farming is a practice which is widely used in the Great Plains of the United States and Canada, in the Columbia Basin wheat area of Washington and Oregon, and in the U.S.S.R. In the United States alone, 37 million acres were reported in cultivated summer fallow in 1964, a year when the total wheat harvested acreage of the country was 48 million acres. The area in cultivated summer fallow has some relationship to the acreage allotment program under which the price of wheat is supported by the Federal Government in return for a commitment from farmers to reduce the acreage which they will plant to wheat. However, the practice of raising a crop of wheat every other year in the high risk subhumid areas was being widely used in the first three decades of this century before the crop allotment program came into existence.

By keeping the land cultivated and free of weed growth during the year when it is not planted to wheat, sufficient moisture can be accumu-

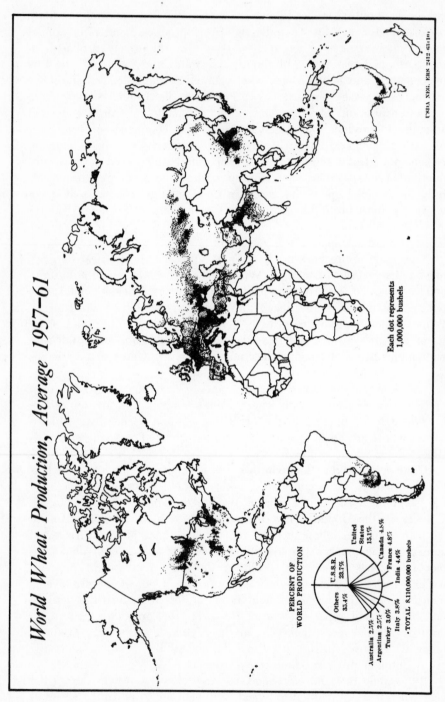

World Wheat Production, Average 1957–61

Each dot represents
1,000,000 bushels

PERCENT OF
WORLD PRODUCTION

U.S.S.R. 23.7%

United States 15.1%

Canada 4.5%

France 4.8%

India 4.4%

Italy 3.8%

Turkey 3.0%

Argentina 2.5%

Australia 2.5%

Others 35.4%

TOTAL 8,110,000,000 bushels

USDA NEG. ERS 2412 63(10)

FIGURE 3.2.

Source: *A Graphic Summary of World Agriculture,* p. 15, U.S. Department of Agriculture, Washington, D. C., 1964.

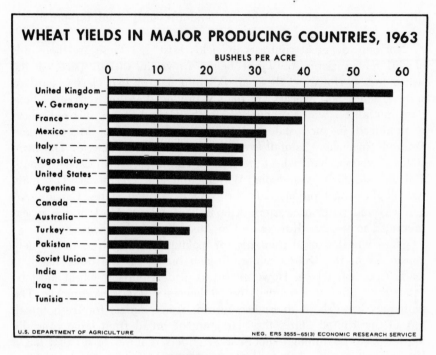

WHEAT YIELDS IN MAJOR PRODUCING COUNTRIES, 1963

BUSHELS PER ACRE

FIGURE 3.3.

Source: L. R. Brown, *Increasing World Food Output: Problems and Prospects*, p. 80, U.S. Department of Agriculture, 1965.

lated to get a better yield of wheat in the year when it is grown. Since there are extensive areas of relatively fertile land available which are suitable mainly for the production of small grains, the practice of cultivated summer fallowing will generally give a better economic return under favorable wheat prices than use of such land for the grazing of cattle. The danger of soil loss through wind erosion, however, can be very great; and probably where such an erosion hazard exists the most reliable use of the land is for the grazing of livestock.

Great variations in wheat yields exist from country to country. In Tunisia, with less than ten bushels per acre, wheat is being produced with relatively primitive methods, little or no fertilization, low yielding, native durum varieties, and low and unreliable rainfall conditions. In contrast, in the United Kingdom where yields were nearly 60 bushels per acre, rainfall is near the ideal, heavy fertilization is used, and carefully selected varieties and other intensive applications of technology are made. Certainly, great increases in world food supplies would result by raising wheat yields to the averages attained in countries having the highest yields. The high rank of Mexico in world wheat yields is a striking example of the increased production that can be realized through improvement of yields on land presently used for producing wheat.

Not only do great contrasts in yields exist, but there are thousands of different varieties of wheat being grown in various parts of the world today. However, these all fall into one of two classes, hard or soft varieties. Soft wheats are mainly produced in areas having abundant rainfall for wheat production. In the United States, soft red wheat is produced in such states as Illinois, Indiana, Ohio, Missouri, and Pennsylvania where rainfall averages about 40 inches a year. The main states producing hard wheat are Kansas, Nebraska, North Dakota, South Dakota, Montana, Minnesota, Washington, and Oklahoma. The main part of the wheat produced in the U.S.S.R., United States, Canada, Argentina, and parts of western Asia is hard wheat. Most of the wheat produced in western Europe and in Australia is soft wheat.

Hard wheat is used primarily for making bread such as that widely consumed in the United States. French bread, on the other hand, is made from soft wheat. However, bread made from soft wheat must be eaten soon after it is made, since it becomes stale very quickly. The numerous breadshops in France furnish ready outlets for fresh bread, but in the United States the marketing of bread from large centrally located bakeries through supermarket outlets requires bread that has a longer shelf life than that which can be made from soft wheat. Most cookies, pastries, cake mixes, rolls, crackers, and other products are made from soft wheat.

Durum wheat is a variety of hard wheat which is used primarily for making spaghetti, noodles, and macaroni. It is grown in the United States, Canada, northern Africa and southwestern Asia, Italy, and the U.S.S.R. In the United States, durum wheat is produced mainly in North Dakota and South Dakota.

rice

Rice is a crop of wet subtropical and tropical regions. Production is overwhelmingly concentrated in south and east Asia, where about 95 percent of the world's rice is produced. The percentage increases in the area used for rice and in production are quite similar to those for wheat as shown by the following tabulation for paddy rice:

	Annual Average 1948-52	Annual Average 1961-65	Percentage Increase 1948-52 to 1961-65
Area (millions of acres)	103	123	19
Production (millions of metric tons)	166	250	51

World Rice Production, Average 1957–61*

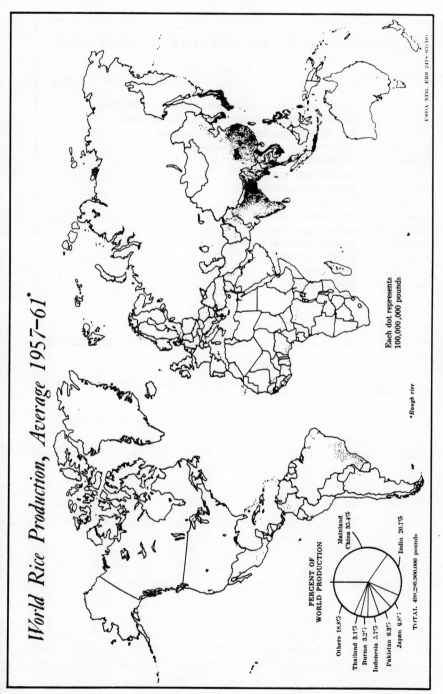

Each dot represents 100,000,000 pounds

*Rough rice

PERCENT OF WORLD PRODUCTION

Mainland China 35.4%

India 20.7%

Japan 6.8%

Pakistan 6.3%

Indonesia 5.7%

Burma 3.2%

Thailand 3.1%

Others 18.8%

TOTAL 498,280,900,000 pounds

USDA NEG. ERS 2418-63(10)

FIGURE 3.4.

Source: *A Graphic Summary of World Agriculture,* p. 22, U.S. Department of Agriculture, 1964.

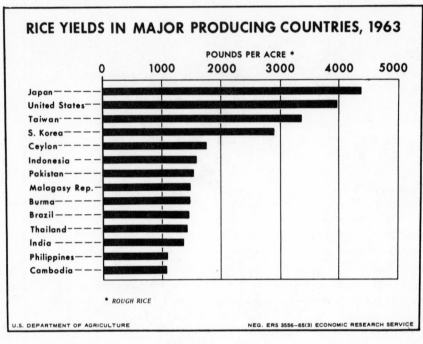

FIGURE 3.5.

Source: L. R. Brown, *Increasing World Food Output: Problems and Prospects.* p. 77, U.S. Department of Agriculture, 1965.

For the countries of southern and eastern Asia, rice has been a basic food for hundreds even thousands of years. Although the place where rice culture originated is not known, it was being produced in China 5,000 years ago. Thus rice has been to the Far Eastern civilizations what wheat has been to the Western World—the staff of life. In the United States, rice has long been a staple in the diet in the southern States. Since neither wheat nor Irish potatoes were suited to the subtropical South, rice was widely used in the antebellum period.

The thousands of varieties of rice can be placed in three main groups: long-grain, medium-grain, and short-grain. Long-grain rice kernels are clear and translucent compared to the chalky appearance of short-grain varieties. The kernels of long-grain are four to five times longer than they are wide. An essential difference in long-grain and short-grain varieties relates to cooking characteristics. After cooking, long-grained varieties stay separated whereas medium-grained and particularly short-grained rice stick together in a gooey mass. Long-grained rice requires a longer growing season than short-grained varieties, thus its production is limited to subtropical and tropical areas. Long-grain rice is the main type of rice produced in southeastern Asia. In Japan and in parts of

China, short-grain rice is produced. Medium-grain rice is the principal kind produced in the United States.

Some rice varieties require as little as three months to produce while others may need from five to seven months to reach maturity. The light requirements for rice production are important and longer days with intense sunlight favor higher yields than shorter days or days with cloud cover or fog. In parts of southern and eastern Asia, considerable rice is grown during the season of low sun and, therefore, needs as many as 240 days to mature compared to about 180 days that is required for summer rice.

Rice requires much more moisture than wheat and very little rice is produced where the annual rainfall is less than 40 inches. Yet too much rain at harvest time can be very damaging; thus, a minimum amount of rice is grown in tropical areas having continuous wetness. Rice is produced both with and without irrigation. Upland or dry rice, which is grown without irrigation, yields much less than paddy or irrigated rice. Yet if there is adequate soil moisture, upland or non-irrigated rice may be produced in hilly areas where irrigation is difficult or impossible and in areas where irrigation water may not be readily available to farmers.

When rice is grown under irrigation, about 15 to 25 inches of water is needed depending on rates of evapotranspiration. During the period of growth the amount of water needed increases to about the middle of the vegetative period and then declines until two or three weeks before harvest when the fields are drained. It is important that a nearly uniform level of about one foot of water be maintained in the fields. This is done by using a careful system of field terraces. In rice areas of the United States, fields are often carefully leveled, which reduces the number of terraces needed. In large rice fields in Arkansas which are also used for producing soybeans, the terraces are reconstructed each year.

Rice is grown on diverse soils; however, an essential characteristic of a good soil for rice culture is a heavy or rather impervious subsoil. Such a condition greatly reduces water loss and also the loss of plant food. Sandy soils or lateritic soils are not good soils for irrigated rice. A major rice-producing area in the United States developed on the Grand Prairie of Arkansas, which prior to the beginning of rice culture about 1900 had been almost completely avoided by cotton farmers because of the poor soil drainage.

Two very different approaches are used in producing paddy or irrigated rice. In southern and eastern Asia, a labor intensive approach is used almost entirely. Great amounts of tedious hand labor are used in producing rice in small, carefully tilled fields of that part of the world. The rice seeds are first sown in a seed bed and are removed to the fields as small plants which are transplanted one by one in standing water. Placing the rice seedlings directly in flooded fields reduces the weed problem considerably. The transplant approach to rice culture has a major advantage in southern and eastern Asia where land is scarce. Because rice remains in the seedbed from 20 to 60 days before transplanting, the rice fields can be multiple-cropped much more easily. Much land in densely populated southern and eastern Asia does produce more than one crop a year.

In parts of the Western World, particularly in the United States, a capital intensive approach is used in the production of rice. On the Grand Prairie of Arkansas of which Stuttgart is the regional center, the author has observed rice being planted and fertilized by airplane and harvested with huge combines similar to those used in harvesting wheat in the Great Plains. Large pumps pump water from the ground water supply and from surface reservoirs which are fed from surface streams and from water pumped from the ground. In recent years the construction of a large number of storage reservoirs has been necessary in order to make more effective use of limited water resources. Very little labor is used in producing rice in the United States compared to the many hours used in producing an acre of rice in southeastern Asia. Rice is generally planted in large fields of 80 acres or more, although in some older rice-growing areas of French Louisiana the fields are relatively small and more labor is used there.

Rice culture in the United States is limited almost entirely to three relatively small areas. These are: (1) the Arkansas Grand Prairie, northeastern Arkansas, with some production on heavy soils of the Mississippi alluvial valley on the Mississippi side of the river; (2) the coastal prairies of Louisiana and Texas; (3) and in the Sacramento Valley of California. Formerly, rice culture was of major importance on the Sea Island coast of Georgia and South Carolina.

Rice yields vary greatly from country to country. It is indeed important to note that Japan and the United States, which have consistently had the highest yields in recent years, are producing rice under rather different circumstances. Although much use of small tractors in Japan in recent years has reduced the labor needed to produce rice, it is still produced in that country on much smaller farms and with more labor than in the United States. If other parts of southern and eastern Asia could match the yields of Japanese farmers, the world's rice supply would be greatly increased in the areas most in need of additional food.

corn

Corn, which is known as maize in most parts of the world, except in the United States, is the third of the world's most basic crops. About 247 million acres produced 226 million metric tons of corn in 1965. This world area compares with 217 million acres being produced on the average from 1948-52. When it is noted that the area in corn in the United States has dropped from 80 to 70 million acres between 1954 and 1967, there has really been a very substantial increase in area being used for corn in other parts of the world. Latin America, southern and eastern Asia, and Africa have had particularly significant increases in area.

The increase in corn production has been even more striking. The average annual production for 1948-52 for the world was 139 million metric tons. The average for the period 1961-65 was 216 million metric tons, which amounts to an increase of 55 percent over 1948-52. Production has increased substantially in all parts of the world, but in Europe, Latin America, Africa, and southern and eastern Asia, the percentage increases have been especially noteworthy. It is important to remember that the greatly increased production has taken place in the United States on several million fewer acres used for corn.

Corn yields in the United States are the highest in the world. In parts of the Midwest corn yields are now averaging more than 100 bushels to the acre. When the author was a boy on a farm in south-central Indiana in the 1930's, yields of 65 bushels were considered exceptionally good. From the chart showing corn yields for leading countries, it is obvious that there is much opportunity for improvement. Yields are improving in some countries and continuing increases will raise production substantially over the next few years.

The many varieties of corn can be grouped into three main categories: dent corn, flint corn, and flour corn. Popcorn and sweet corn which are well-known in the United States are relatively minor varieties in other parts of the world. Europeans generally do not even know what corn-on-the-cob is. Dent corn has a depression in the crown of the kernel, which is related to the presence of soft starch in the kernel. Yellow dent varieties of corn are grown mainly in the United States and in northern Mexico. Some white dent corn is also grown in these countries. Flint corn is produced primarily in Latin America, Europe, and Asia. This type of corn mainly has a hard starch content. There is less moisture in flint corn than in dent corn; therefore it dries easily and stores well. Flour corn has a high soft starch content and is grown particularly in Colombia, Peru, and Bolivia.

Corn is grown under more contrasting climatic conditions than any of the other cereal grains. It is grown as far north as the 58° parallel

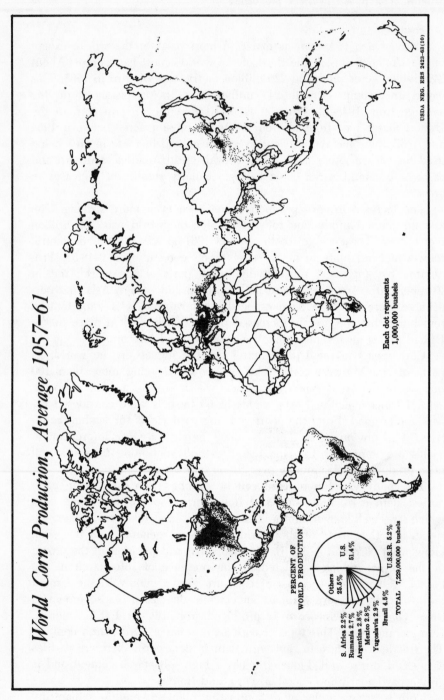

World Corn Production, Average 1957–61

PERCENT OF
WORLD PRODUCTION

Others
25.5%

U.S.
51.4%

S. Africa 2.2%
Rumania 2.7%
Argentina 2.8%
Mexico 2.8%
Yugoslavia 2.9%
Brazil 4.5%
U.S.S.R. 5.2%

TOTAL 7,229,000,000 bushels

Each dot represents
1,000,000 bushels

USDA NEG. ERS 2423-63(10)

FIGURE 3.6.

Source: *A Graphic Summary of World Agriculture,* p. 27, U.S. Department of Agriculture, 1964.

in Canada and the Soviet Union and in the southern hemisphere it reaches the 40° parallel. It is grown below sea level around the Caspian Sea and above 12,000 feet in the Andes of Peru. No other cereal crop is so widely distributed, although wheat is planted on a greater area. Corn or maize originated in the Americas and its spread to other parts of the world has occurred since it was first discovered among the American Indians 475 years ago.

Essentially, corn likes warm weather and plenty of moisture. Very little corn is produced in the United States where the average temperature for the three summer months is below 66° F. and where night temperatures average less than 55° during the summer. A frost-free period of about 140 days is required if corn is to be harvested as grain rather than as silage. Some hybrid varieties have been developed in recent years which can be matured in a somewhat shorter time span. Not only is it essential that an abundant supply of moisture is available for best yields, it is important also that this supply be favorably distributed during the growing season. A particularly crucial time when much moisture is needed comes when the corn is silking and tasseling. Some corn is now being irrigated in order to assure that adequate

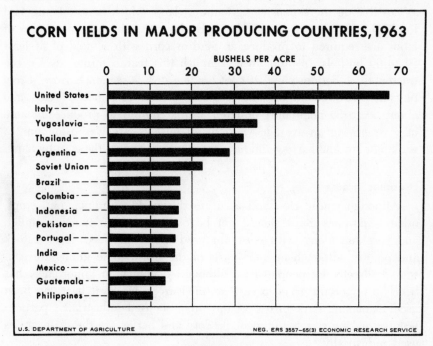

FIGURE 3.7.

Source: L. R. Brown, *Increasing World Food Output: Problems and Prospects*, p. 83, U. S. Department of Agriculture, 1965.

moisture is available at critical periods. However, as in the case of wheat and rice, corn is being grown extensively in many parts of the world where physical conditions for its production are not ideal.

As was true for wheat and rice, many different combinations of land, labor, and capital are used in the production of corn. In some parts of the world such as in the Andean area of South America, corn is being produced in small terraced fields on steeply sloping land. These fields are laboriously cultivated by the native Indians who used only a primitive technology that includes no use of tractor power and no mechanical planters, cultivators, and harvesters. In the midwestern United States, corn is produced in large fields that dominate the relatively flat till plains left by glaciation. Soils formed under a prairie grass vegetation have been excellent for corn. In many instances, corn may be planted in the same fields continuously but most farmers prefer to rotate corn with soybeans and other crops such as oats and wheat, which generally serve as a nurse crop for legumes such as red clover and alfalfa.

Here in the heartland of American agriculture, corn is king and the most modern techniques are used to produce as much corn as efficiently as possible on the expensive agricultural land, which is now being sold for $700 to $1,000 per acre. Today only about five to seven man-hours of labor are required to produce an acre of corn with a yield of at least 80 to 90 bushels. In order to accomplish this feat, a farmer uses a big tractor, three-bottom plow, 10-foot tandem disk, 4-section harrow, 4-row planter, 4-row cultivator, and a 2-row picker. A hundred years ago about 30 to 35 man-hours of labor were required to produce an acre of corn having an average yield of 40 bushels. He was using only a walking plow and harrow and he planted and harvested the corn by hand.

other grains

Although wheat, rice, and corn are the world's leading grain crops, other important grains should not be overlooked. Barley, oats, millet, and sorghum along with corn are used mainly as feed for livestock and poultry, although in some parts of the world these grains are consumed directly by people. Rye, although declining in importance, has been an important grain in east-central Europe where it has long been used in making dark bread. Grains are also used industrially, particularly in making starch and alcohol. Corn and barley are the main grains used industrially.

The world acreage and production of these other grains are summarized in the following tabulation.

	Million acres (1965)	Million metric tons (1965)
Oats	76	47
Barley	171	105
Rye	67	35
Millet and Sorghum	270	78

Barley is grown generally in the same parts of the world as wheat but it is grown farther north than wheat in Europe and under more arid conditions in northern Africa and America. Oats are mainly grown in the United States, Canada, Europe, and the U.S.S.R. Rye is grown primarily in central and eastern Europe and in the Soviet Union. Sorghum and millet, which are used for food in Asia and Africa and for feed in most other parts of the world are particularly important in the northern part of mainland China, India, and the southern Great Plains of the United States.

vegetable oils

Soybeans and peanuts (groundnuts) are the world's leading sources of vegetable oil; however, there are a number of other vegetable oils that are particularly important in some parts of the world. Some vegetable oils enter into world trade much more than do others. The leading vegetable oils and the percentage of total world production supplied by each kind for the period 1957-61 are as follows:

	Percent
Soybean	18.6
Peanut	12.8
Coconut	12.1
Cottonseed	11.7
Palm and palm kernel	10.0
Sunflower seed	9.0
Olive	6.9
Other	18.9

Important producing countries of the major vegetable oil crops in 1965, listed in order of importance are as follows:

Soybeans—United States and Mainland China

Peanuts—India, Mainland China, Nigeria, Senegal, and United States

Coconuts—Philippines, Indonesia, India, and Ceylon

Cottonseed—United States, U.S.S.R., India, and Brazil

Palm and palm kernel—Nigeria, Republic of Congo, Indonesia, West Malaysia, and Brazil

Sunflower seed—U.S.S.R., Argentina, and Romania
Olive—Italy, Spain, Greece, and Portugal
Flax (linseed oil)—United States, Canada, Argentina, India, and
. U.S.S.R.
Rapeseed—India, Mainland China
Sesame seed—India, Mainland China
Castor beans—Brazil, India
Hempseed—U.S.S.R.

Soybeans, the world's leading vegetable-oil crop, have been particularly important in eastern Asia and more recently have become a very important crop in the United States. The soybean was introduced into the United States from eastern Asia during the last century, but it did not really become a major crop until Federal programs for regulating the production of such crops as cotton, corn, and wheat forced farmers to search for another good cash crop. Also very important was the major switch from animal to vegetable fats that has occurred in American and European diets. In eastern Asia the soybean is used in a great variety of ways as a major food for human consumption. Soybean oil is the main product of soybeans grown in the United States. This oil is used for making oleomargarine, vegetable cooking compounds, soap, and paint.

Since the climatic and soil requirements for producing soybeans are quite similar to those needed for corn, this has become an ideal crop for farmers to use as an alternate source of income in the midwestern United States. Similarly, soybeans are well suited to many of the areas where cotton is or has been a major crop in the United States. The combine which is used to harvest soybeans is similar to that used for wheat and other small grains. In recent years, farm machinery manufacturing companies have developed a machine that can be used for picking corn as well as for harvesting soybeans if certain basic parts are interchanged. This results in saving a considerable capital investment in additional harvesting equipment.

Peanuts or groundnuts are produced mainly in tropical and subtropical parts of the world but in some places they have also been produced in mid-latitude locations. The peanut plant needs warm weather during the growth period. The best seasonal distribution of precipitation provides a good moisture supply when the nuts are developing, followed by drier weather and plenty of sunshine during the harvest period. A loamy sand or a sandy loam soil is the most favorable soil for peanuts. Heavy clay soils are very unfavorable because the peanut fruit is unable to penetrate into such heavy soils effectively.

The major producing areas are in India, west Africa, China, and the United States. In the United States, the two principal concentrations

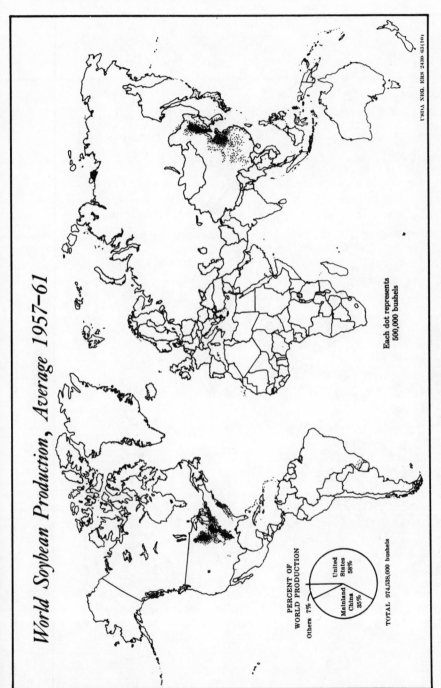

World Soybean Production, Average 1957–61

Each dot represents
500,000 bushels

**PERCENT OF
WORLD PRODUCTION**

Others 7%

United
States
58%

Mainland
China
35%

TOTAL 974,538,000 bushels

USDA NEG. ERS 2439-64(10)

FIGURE 3.8.

Source: A Graphic Summary of World Agriculture, p. 34, U.S. Department of Agriculture, 1964.

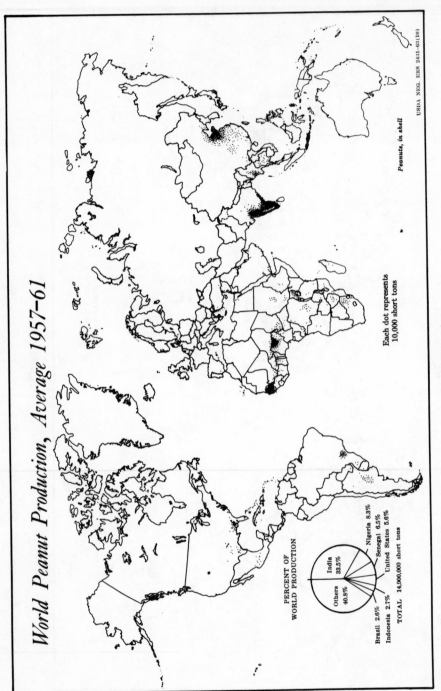

World Peanut Production, Average 1957–61

PERCENT OF
WORLD PRODUCTION

India
33.5%

Others
40.8%

Nigeria 8.3%

Senegal 6.5%

United States 5.6%

Brazil 2.6%

Indonesia 2.7%

TOTAL 14,000,000 short tons

Each dot represents
10,000 short tons

Peanuts, in shell

USDA NEG. ERS 2431–63(10)

FIGURE 3.9.

Source: A Graphic Summary of World Agriculture, p. 35, U.S. Department of Agriculture, 1964.

of peanut production are located on the southeastern coastal plain: one in northeastern North Carolina and southeastern Virginia and the other in southwestern Georgia, southeastern Alabama, and northern Florida. A third area of lesser importance is located in eastern Texas and Oklahoma. Following the boll weevil infestation of cotton producing areas in southeastern Alabama, farmers began to plant peanuts and raise hogs in order to have an alternate source of income. This diversification of crop and livestock production actually resulted in raising farm incomes in this part of Alabama. Today, the visitor to Enterprise, Alabama, in Coffee County will see a monument to the boll weevil on the main street which was erected in gratitude to a pest that caused them to change their system of farming in the period from 1915 to 1925.

cotton

In American history, cotton culture provided the basis for great contrast in economic development during the past colonial period. Even after the Civil War the dominance of a one-crop agricultural economy in the South continued to play a major role in shaping the socio-economic structure of this part of the nation. The peak acreage of cotton was reached in 1926 when 44.6 million acres were harvested. Following the rapid introduction of synthetic fibers in the United States such as rayon, orlon, dacron, and others, the domestic need for cotton fiber began to decline. Coincident with this competition from synthetic fibers, cotton production in the United States began to face serious competition from developing cotton production in such countries as Mexico and Brazil and from the longstanding production of India where cotton was being produced with labor that could be hired much more cheaply than the former American Negro slaves who were beginning to find many new opportunities for employment in northern cities following the passage of strict immigration laws in the early 1920's.

The following brief tabulation tells a significant story about cotton production in the United States during the past 100 years:

Year	Acreage Harvested (million acres)	Yield Per Harvested Acre (pounds)
1866	7.7	122
1900	24.9	195
1925	44.4	174
1950	17.8	269
1959	14.6	461
1964	13.9	530

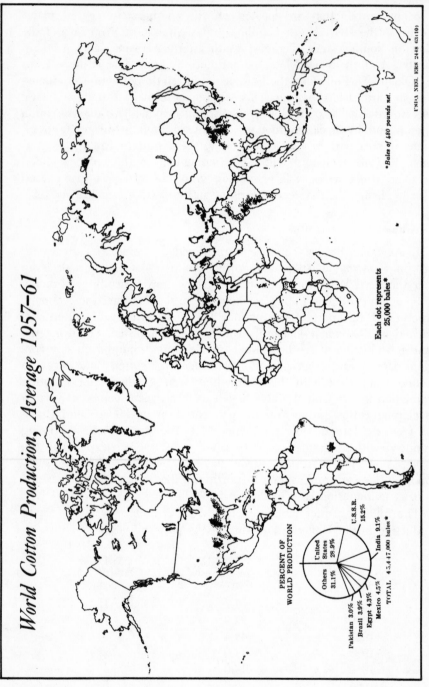

World Cotton Production, Average 1957–61

Each dot represents
25,000 bales*

PERCENT OF
WORLD PRODUCTION

United
States
28.9%

Others
31.1%

U.S.S.R.
15.2%

India 9.1%

TOTAL 45,447,000 bales*

Pakistan 3.0%
Brazil 3.9%
Egypt 4.3%
Mexico 4.5%

*Bales of 480 pounds net.

USDA NEG. ERS 2448 (6-3410)

FIGURE 3.10.

Source: A Graphic Summary of World Agriculture, p. 52, U.S. Department of Agriculture, 1964.

64

Not only has this remarkable change in total acreage and increase in yield per acre occurred, but also of major significance is the fact that a strong westward migration of cotton production has taken place. In 1964 the Mississippi Delta and Southern Plains states along with California, Arizona, and New Mexico accounted for 76 percent of the total acreage and production of cotton harvested. In 1900 these states accounted for only 57 percent of the total acreage harvested.

In addition to a frost-free season of 180 to 200 days, cotton needs a mean annual temperature of at least 50° F., and 60° F. is preferable. Rainfall for the year should be at least 20 inches with suitable concentration of rainfall during the growing season. As much as 60 to 75 inches of rain would be satisfactory if there is a relatively dry harvest period. Plenty of sunshine and hot summers are also basic physical requirements for cotton culture.

Cotton is grown commercially in more than 75 countries under many different soil and climatic conditions and with great contrasts in the technology used. However, twelve countries account for about nine-tenths of the total world production. These are the United States, the Soviet Union, Mainland China, India, Mexico, Brazil, the United Arab Republic, Pakistan, Turkey, the Sudan, Syria, and Peru.

sugar

Main sources of the world's sugar supply are sugar cane and sugar beets. Sugar cane is a widely produced crop in the tropics and subtropics and sugar beets are almost entirely produced in the mid-latitude areas of the northern hemisphere. Since sugar cane is a perennial, the first crop, which is harvested after about 15 to 24 months, is followed by at least two *ratoon* crops, which mature at intervals of about one year. In subtropical areas such as Louisiana and Florida, frost is a real hazard and often the cane is cut before reaching maturity, which means that there is a relatively low sugar content at the time of cutting. In contrast, the sugar beet, which now supplies about two-fifths of the world's certrifugal sugar is capable of being grown in climates with cooler summers and a shorter growing season. Diseases are a common threat to both sugar cane and sugar beets.

Government controls on production, refining, and marketing of sugar have an important influence in determining how much sugar is produced and where. In most countries where sugar is produced, a minimum price level to growers is guaranteed in one way or another. Countries which import sugar generally levy import duties or use other control measures, particularly if there is a domestic sugar industry that needs protection. Prior to 1960 about 95 percent of the sugar imported

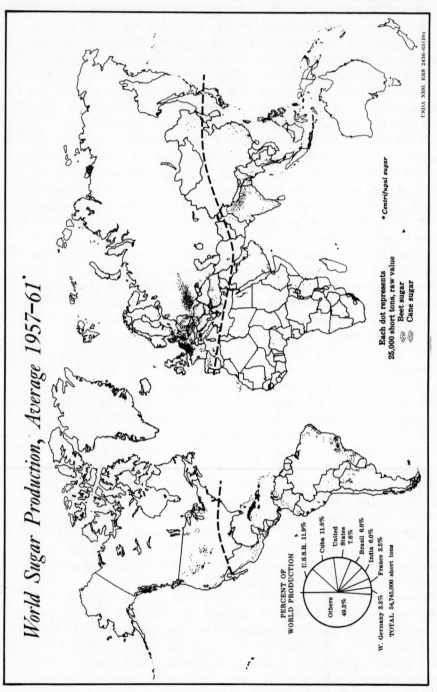

*World Sugar Production, Average 1957–61**

Each dot represents
25,000 short tons, raw value

Beet sugar
Cane sugar

*Centrifugal sugar

PERCENT OF
WORLD PRODUCTION

U.S.S.R. 11.9%
Cuba 11.8%
United States 7.6%
Brazil 6.6%
India 6.0%
France 3.5%
W. Germany 3.3%
Others 49.3%

TOTAL 54,745,000 short tons

USDA NEG. ERS 2436-63(10)

FIGURE 3.11.

Source: A Graphic Summary of World Agriculture, p. 40, U.S. Department of Agriculture, 1964.

by the United States was supplied by Cuba and the Philippines. By Congressional action the number of countries which have been assigned quotas to sell sugar to the United States has been increased to 25. The United Kingdom and other European countries also have preferential arrangements with various producing countries.

potatoes

Europe and the U.S.S.R. produce about three-fourths of the world's potato crop. These potatoes are used as food, are fed to livestock, and are used to make industrial alcohol. Potatoes are also produced in Andean America, the central highlands of Mexico, eastern Brazil, on the Pampa of Argentina, in the Ganges Valley of India, Japan, parts of Africa, and in the United States. In this country, Maine and Idaho are the leading potato states but North Dakota, Minnesota, California, New York, Colorado, Wisconsin, and Michigan are other important producers.

The potato was originally grown in Andean America, where it still remains a basic food crop for the native Indian people. The crop is well suited to areas with cool summers that are moderately humid. High temperatures are unsatisfactory for potato production since no tubers are produced under such conditions. In the United States, early potatoes are grown in California and Florida in the cool part of the year and shipped north in advance of maturing of the potato crop in more northern commercial areas.

coffee and tea

Coffee is predominantly a crop of Latin America, which produces 69 percent of the world's coffee; whereas tea is a product mainly of Asia, which supplied nearly 90 percent of the tea produced in the world in 1965. Brazil alone produces nearly half of the world's coffee with Colombia second with only about 12 percent of the world's production. Mexico, Angola, and the Ivory Coast are other important coffee-producing countries. Many other countries produce some coffee. India, Ceylon, and Mainland China are the dominant producers of tea. Japan, Indonesia, and Pakistan are other tea-producing countries in Asia. Some tea is produced in Africa and the U.S.S.R.

The main species of coffee are Arabica and Robusta with Arabica accounting for about four-fifths of the world's production. Nearly all of the coffee produced in Latin America is Arabica while most of that grown in Africa is Robusta. Coffee is grown both with and without shade. Most of the coffee in Brazil is grown without shade. Coffee is also produced over a wide range of soils and elevation in the tropics. Moisture requirements for coffee are moderate for a tropical crop. The total annual rainfall of the coffee-producing areas generally ranges from 50 to 90 inches. A minimum of rain during the flowering season is very important. The life of the coffee tree is about 30 to 40 years with trees beginning to bear at about three to five years of age.

There are two main types of tea, the Assam tea and Chinese tea. Tea is a plant of the subtropics and it could be produced in many areas such as southeastern United States where it is not being produced now. Year-round warmth and moisture is the most desirable climatic condition for tea production. As in the production of coffee, tea is produced on a wide variety of soils and at a wide altitudinal range.

tobacco

Originally grown in the Americas, tobacco is now widely grown from the shores of Lake Huron in Canada to the island of Sumatra in the tropics. The United States and Mainland China are the two leading tobacco-producing countries but it is also grown in Europe, southwestern Asia, Latin America, and in southern and eastern Asia in significant amounts. Many different methods of curing and marketing tobacco are used.

In the United States the acreage of tobacco harvested has ranged between one and two million acres since 1900 except for the year 1930, when a peak acreage of 2.1 million acres was reported. North Carolina and Kentucky are leading producing States and together account for about three-fifths of the total acreage. Virginia, South Carolina, Tennessee, and Georgia in that order are the next four ranking States. During the past thirty years, the acreage of tobacco has changed very little, mainly because an attempt is being made to keep supply in line with demand under an acreage allotment program, which provides for a price guarantee for participating farmers.

hay and forage crops

A great variety of grasses and legumes are grown as hay and forage crops under a wide variety of physical conditions. It will not be possible to describe all of these here, but the pattern of production in the United States is analyzed briefly. Generally, in this country the acreage used for hay crops has fluctuated between 60 and 70 million acres since 1900. Two main types of hay are cut: tame hay and wild hay. The principal tame hay crops are alfalfa, clover, timothy, small grains cut for hay, and lespedeza. Most of the wild hay is cut in the northern Great Plains where selected areas of native grasses, which are also used for grazing, are cut for hay.

The most widely grown hay crop is alfalfa and alfalfa mixtures. The only major area in which alfalfa is of little importance is in the southeastern States, where a humid climate and sandy soils are not conducive to its production. Soils with adequate lime are the most favorable soils for growing alfalfa. In the western States, it is an irrigated crop. It has been widely used in irrigated areas to build up organic matter in soils which formed in semiarid and arid climates and had very little natural organic matter. In the northern Great Plains, a considerable acreage of alfalfa is grown without irrigation. It is grown not only for hay but also for seed. Hardy varieties grown in these States are not so easily damaged by winter killing as are varieties grown in warmer areas. The largest concentration of alfalfa acreage is in the southern part of the Lake States and the northern part of the Corn Belt where soils favorable for its production coincide with areas in which dairying has become the major type of farming.

Clover and timothy are often planted as a mixture. Most of the timothy and clover cut for hay is grown in the north-central and northeastern States. It is still the major hay crop on many soils that are not suited to production of the higher yielding and better quality alfalfa hay. Timothy and clover mixture as a hay crop is not as expensive to seed and is less likely to suffer damage from winter killing than is alfalfa. Lespedeza is a comparatively new crop among the hay and forage crops in widespread use in the United States. As a legume it has found ready acceptance in the mid-South where soils are not too favorable for the production of alfalfa. Because of the milder climate, which makes it possible to graze animals longer than in more northern areas, much less hay and forage crops are grown in that part of the country. Feeding of silage stored in upright or trench silos is a popular feed for both dairy and beef cattle.

vegetable crops

A great variety of vegetable crops are grown under very different physical and economic circumstances. In some parts of the world, vegetables are consumed mainly on the farms on which they are grown. In other countries having sizable urban populations, market gardens located relatively near to large urban centers produce vegetables for the fresh market. In still other parts of the world processing of vegetables by canning and quick freezing permits the use of vegetables when fresh vegetables are not available. In still other situations such as the United States, fresh vegetables can be produced in warm subtropical areas during the winter and shipped by fast refrigerated trucks and rail cars to markets located in more northern areas.

Although many different vegetables are produced in the United States, the ten leading vegetables in order of the acreage harvested in

1964 and with thousands of acres shown in parenthesis are: sweet corn (546), green peas (395), tomatoes (389), snap beans (280), watermelons (246), lettuce (210), asparagus (139), cantaloupe (124), cucumbers and pickles (110), and cabbage (106). Four relatively small parts of the country have a particularly high dollar value of vegetables harvested for sale: (1) the irrigated areas of California, including parts of the Central Valley, the Imperial Valley, and the Santa Clara and other coastal valleys, some of these areas producing during late fall, winter, and early spring; (2) the south Florida vegetable areas where most of the production takes place during late fall, winter, and early spring when areas farther north are unable to produce vegetables; (3) the lower Rio Grande Valley of Texas which also produces vegetables during the off seasons for northern parts of the country; and (4) the Middle Atlantic Coastal Plain which produces both for processing and for fresh market.

fruits and nuts

As is true of vegetables, there is a great variety of fruits and nuts being produced under greatly contrasting physical and economic circumstances in many parts of the world. A brief examination of the production pattern in the United States will serve to illustrate this diversity.

One category of fruits is that of berries of which strawberries, blueberries, cranberries, raspberries, and blackberries were the leading kinds by value produced in 1964 in the United States. The value of strawberries produced was 97 million dollars which comprised about 67 percent of the total value of all berries produced.

Tree fruits constitute the main group of fruits and, along with grapes and nuts, accounted for a total value of more than 1.5 billion dollars produced in the United States in 1964. The ten leading fruits produced in the United States in 1964 with the value of production indicated in parentheses in millions of dollars are as follows: oranges (455), apples (222), grapes (215), peaches (138), grapefruit (85), plums and prunes (69), pears (62), lemons (54), cherries (51), and apricots (26). The leading kinds of nuts produced are: English walnuts (36), almonds (36), and pecans (14).

Fruits bring a relatively high return per acre and their production is highly localized. To a major degree, climatic conditions play an important role in the selection of areas for fruit production. California and Florida are the leading fruit-producing States. About two-thirds of the total value of all fruits and nuts sold is contributed by these two States. Both citrus and deciduous fruits are of major importance in California as well as nuts, particularly walnuts and almonds. In Florida, citrus fruits dominate, although some small fruits, such as strawberries, and some pecans are produced. Other important fruit-producing areas contributing a high value of fruits are the irrigated valleys of Washington and Oregon, where apples and pears are especially important; the eastern and southern shores of the Great Lakes, having grapes, cherries, apples, and peaches; and the valley slopes of Virginia and Maryland where apples and peaches are grown.

other crops

Production patterns of numerous crops have not been discussed because space does not permit adequate attention in this book. However, brief mention of these is made here in order that their importance to some parts of the world can at least be recognized. Among the food crops that have not been discussed, sweet potatoes, yams, yuca, manioc (cassava), and a great variety of pulses are particularly important in tropical and subtropical parts of the world. Many fruits, nuts, and vegetables have not been mentioned. The use of grapes for making wine has not been discussed. Cacao from which cocoa and chocolate is made is a significant crop produced in the wet tropics of West Africa and Latin America. The use of flax for fiber and other important fiber crops such as hemp, jute, sisal, henequen, abacca, and others have likewise been omitted. The localization of some of these fibers provides opportunity for worthwhile geographical analysis. Natural rubber, which is particularly concentrated in southeastern Asia, is still another commodity generally considered to be a product of agriculture. References cited at the end of this chapter will provide additional information about such crops.

Patterns of Livestock Production

Cattle, swine (hogs), sheep, goats, and poultry are the most widely distributed livestock classes. Horses, mules, asses, camels, and buffalo are raised primarily for draft purposes rather than as a source of food. Meat, milk, eggs, and wool are the main livestock products. Altogether, these livestock products account for about two-fifths of the total value of world agricultural production. Meat represents nearly three-fifths of the total value of livestock products, milk accounts for more than a fourth, eggs about a tenth, and wool about one-twentieth. With the exception of wool, the production of livestock products is especially concentrated in the United States and Western Europe. Together, these two regions produce about two-fifths of all livestock products now being produced in the world.

cattle

The cattle population of the world exceeded a billion animals in 1965. India has a particularly dense cattle population. Partly because of religious beliefs, the present cattle population of that country contrasts greatly with the amount of meat and milk derived from the cattle population of that country. In the United States, which has the second largest cattle population, about a fifth of the animals can be classified as being used primarily for dairy purposes while more than four-fifths are principally raised for beef production. Brazil, U.S.S.R., Argentina, and Pakistan are other leading countries in cattle population.

Cattle vary greatly in quality from one part of the world to another. The output of meat per animal in Europe is about ten times as great as that in eastern Asia and seven times as great as that for Africa. This great difference in animal productivity is mainly due to poor management, including little attention to breeding practices, control of diseases and insects, and provision of proper feed found particularly in much of Asia and Africa. Thus if the meat and milk production of these areas is to be increased, first attention must be given to increasing the productivity per animal rather than to increasing the number of animals.

The world's cattle population can be appropriately divided into two main kinds: European and Indian (also known as zebu). The main European breeds are: Friesian-Holstein, Guernsey, Jersey, Ayrshire, Red Danish, and Brown Swiss which are used as dairy cattle; and Herefords and Aberdeen Angus which are beef breeds. The Shorthorn is used both for milk and beef production. The United States, Canada, Australia, and New Zealand have cattle which are directly descended from those north European breeds. In Latin America, early cattle brought in by the Spanish and Portuguese have later been crossbred with modern European breeds particularly in Argentina, Uruguay, and southern Brazil. In the more tropical areas of Latin America, crosses with Indian cattle have been made. In southern and eastern Asia, Indian cattle are the main type. These animals are much more tolerant of heat and humidity conditions in the wet tropics than are European breeds but they generally have been inferior producers of meat and milk.

hogs

Four main clusters of hogs may be noted on the accompanying map: the Corn Belt of the United States, Mainland China, north central Europe, and eastern Brazil. In these four main areas of production, great contrasts exist in the types of feed used, the number of hogs raised per farm, and the methods used in producing hogs. Hogs are much more efficient animals than cattle are in converting feed into

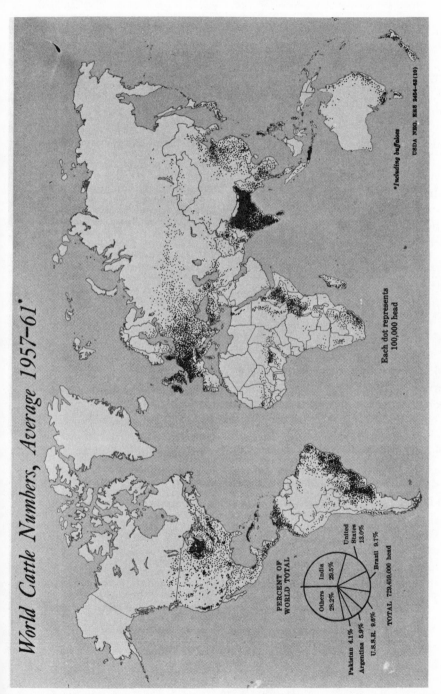

World Cattle Numbers, Average 1957–61[*]

Each dot represents
100,000 head

PERCENT OF
WORLD TOTAL

India 29.5%

United
States 13.0%

Brazil 9.7%

Others 28.2%

Pakistan 4.1%
Argentina 5.9%
U.S.S.R. 9.6%

TOTAL 729,650,000 head

*Including buffaloes

USDA NEG. ERS 2464–63(10)

FIGURE 3.12.

Source: A Graphic Summary of World Agriculture, p. 58, U.S. Department of Agriculture, 1964.

meat. Hogs also reproduce more rapidly than cattle. A sow will have two litters of six to twelve pigs each during a year. Whereas a cow will have only one calf during about the same period of time.

In the United States, corn is the main feed used in fattening hogs. In Europe, barley, rye, or potatoes fed sometimes in combination with skim milk and whey are the main hog feeds. In China, hogs are often fed leftovers from human consumption and even in the United States, around some large urban centers, hogs are fattened on garbage. Hogs are not produced in areas where the Mohammedan and Hindu religions are prevalent.

In the United States a lard-type hog was developed when there was a strong demand for lard as a cooking fat. In Europe the main type of hog is the bacon type. Barley, rye, potatoes, and other root crops fed to hogs in Europe provide a lower fat-content diet than corn which has been the main feed for hogs raised in the United States. With a growing demand for leaner bacon and more lean meat from hogs here in the United States, breeders are hard at work trying to develop a hog to satisfy the American consumer and at the same time one that can utilize corn as its basic feed.

sheep and goats

Distribution of the world's sheep population has some interesting contrasts with that for cattle and hogs. In southwestern Asia, sheep are much more important than either hogs or cattle. In Australia, New Zealand, and South Africa, sheep are of great importance whereas the hog population is very sparse. In southern Europe, sheep are more important in most areas than hogs and much more important in some areas than cattle. In southern Argentina there is a heavy concentration of sheep where practically no cattle or hogs are raised. On the other hand, sheep are practically of no importance in eastern and southeastern Asia.

Goats are of major importance in parts of southern Europe, southwestern Asia, India, Pakistan, Java, western and eastern Africa, parts of southern Africa, northeastern Brazil, Andean America, Mexico, and in areas of nomadic grazing in Mongolia and western parts of Mainland China. By country, India has about seventeen percent of the world's goat population, followed by Turkey with seven percent, Ethiopia with five percent, Iran with four percent, and Brazil with three percent.

A very important reason for the present distribution of sheep and goats relates to the willingness of these animals (particularly goats) to eat grasses, shrubs, and rough forage that cattle will not eat. Thus, in humid areas, these animals use rough grazing land not suitable for other purposes. In semiarid and arid parts of the world, sheep and

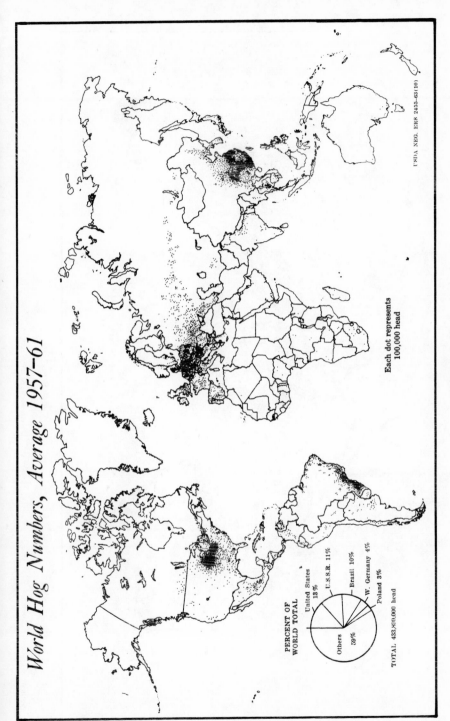

World Hog Numbers, Average 1957–61

USDA NEG. ERS 2455–63(10)

Each dot represents
100,000 head

PERCENT OF
WORLD TOTAL

United States 13%
U.S.S.R. 11%
Brazil 10%
W. Germany 4%
Poland 3%
Others 59%

TOTAL 433,889,000 head

FIGURE 3.13.

Source: *A Graphic Summary of World Agriculture*, p. 59, U.S. Department of Agriculture, 1964.

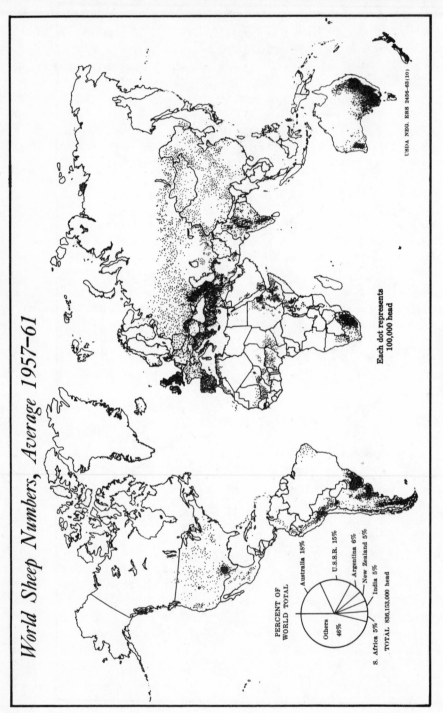

World Sheep Numbers, Average 1957–61

Each dot represents
100,000 head

USDA NEG. ERS 2455–63(10)

PERCENT OF
WORLD TOTAL

Australia 18%
U.S.S.R. 15%
Argentina 6%
New Zealand 5%
India 5%
S. Africa 5%
Others 46%

TOTAL 886,153,000 head

FIGURE 3.14.

Source: *A Graphic Summary of World Agriculture,* p. 60, U.S. Department of Agriculture, 1964.

goats will graze and browse on native plants that cattle will not eat. Furthermore, they are able to exist on less water and to subsist longer without having water. For this reason sheep are found in large numbers in cool, humid parts of the world such as the British Isles and on the margins of some of the world's great deserts such as the Australian, Arabian, Sahara, and Gobi.

Many different breeds of sheep have been developed but only a few are of major significance. Sheep breeds that are known mainly for the production of meat (mutton) are not of great value as wool producers. Wool breeds, such as the well-known Merino, have relatively little value for meat production. In some parts of the world, dual purpose breeds are used to produce saleable mutton and wool. The Merino and breeds developed from the Merino were particularly important in Australia, the Pampa and Patagonian areas of Argentina, New Zealand, western North America, and in southern Africa during the period of early agricultural development in those parts of the world. Today, this breed is of little importance in New Zealand, Argentina, and North America where more emphasis is placed on the production of meat with wool being a by-product of the meat production.

poultry
Poultry is a group of fowl which has been domesticated for meat and egg production. The main types of poultry are chickens, turkeys, ducks, geese, pigeons, bantams, guineas, and pheasants. Commercial poultry production has been of increasing importance in the United States and Europe as a source of food in recent years. The consumption of poultry meat in the United States has increased from about 15 pounds per person in 1935 to about 45 pounds in 1967. This growing popularity is in large part due to lower prices for poultry meat and eggs relative to prices for beef and pork. Improvements in the efficiency of feed conversion have led to more economical egg and poultry meat production in recent years.

The distribution of broiler production in the United States is characterized by a high degree of very heavy concentration within localized areas of several states in the southern and northeastern parts of the country. The increase in mass production of broilers as a highly specialized enterprise is one of the striking changes that has been taking place in the supply of poultry meat. This large-scale production of broilers is particularly important on many small farms which have become uneconomical in size for the production of such crops as cotton and for the production of other livestock products. The organization of loaning money, of feed distribution, and of the dressing and marketing of broilers by large companies in areas having many small capital-

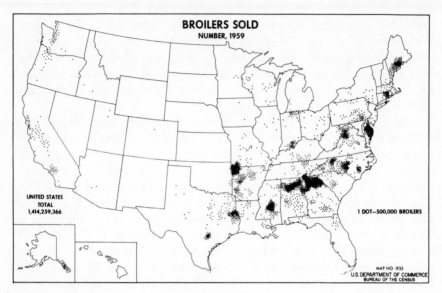

FIGURE 3.15.

deficient farms accounts for many of the main areas of broiler production. In Delaware and on the eastern shore of Maryland and Virginia, a striking concentration of broiler production developed close
to large urban markets in the Northeast before quick-freezing techniques and rapid refrigerated truck transportation was available.

A similar concentration of chicken egg production has also occurred
in the United States in recent years. Similar forces have operated to
bring about the increasing localization of the egg industry, but the
areas that are most important for egg production differ somewhat in
location from those specializing in broilers.

REFERENCES

Farmer's World: The Yearbook of Agriculture, 1964, Washington, D. C.:
    U.S. Department of Agriculture.
Production Yearbook, 1966, Rome: Food and Agriculture Organization of the
    United Nations, 1967.
The State of Food and Agriculture, 1967, Rome: Food and Agriculture Or-
    ganization of the United Nations, 1967.
United States Census of Agriculture, 1964: General Report, Vol. 2, Wash-
    ington, D. C.: Bureau of the Census, U.S. Department of Commerce, 1967.
United States Census of Agriculture, 1964: Maps, Vol. 3, Part 5, Washington,
    D. C.: Bureau of the Census, U.S. Department of Commerce, 1969.
VAN ROYEN, WILLIAM. *The Agricultural Resources of the World.* New York:
    Prentice-Hall, Inc., 1954.

Systems of
Agricultural Production

Production, marketing, and use of the world's diverse agricultural commodities fit together to form regional patterns that can be identified and classified into systems or types of farming that are meaningful. Such a classification adds another dimension to the analysis of agricultural production, since it provides an opportunity to study areal relationships that exist in producing several or single agricultural commodities under a particular combination of physical, historical, economic, social, political, and technological circumstances. The principal objective of this chapter is to provide a framework for further study, since it is not possible within the limits of this book to describe and analyze in detail the many systems and sub-systems which are recognizable.

The study of systems of agricultural production offers a helpful approach to obtaining a more complete understanding of the problems of agriculture in the Worlds of Plenty and Poverty. The composite circumstances that contribute to the existing problems facing agricultural activities today have a space and time perspective that must be appreciated. The village-based peasant farmer with his small tract of land in India has a different set of problems than the *ejido* farmer of Mexico or of the sharecropper of an earlier period in the southern United States. They all share certain common problems but the historical, social, political, and other conditions that have contributed to their relative impoverishment are different in many ways. Systems of agriculture generally change slowly, but in some parts of the world, changes occur more rapidly than in others. The tenure arrangement involving the use of sharecroppers in place of slaves on the southern plantation evolved rapidly after slavery was abolished. The more recent demise of the sharecropper system is occurring over a span of twenty to thirty

years after the initial accelerated migration of the Negro farmer to
urban centers followed World War II. However, in many parts of the
world the peasant farmer still tills the soil with primitive tools and
following traditional methods that have been used over long periods
of time.

Characteristics of Agricultural Systems

The first and still foremost attempt made by an American geographer
to develop a classification of systems of agriculture was made by Der-
went Whittlesey in the mid-thirties after many years of work on this
problem with Wellington Jones. In 1936 he published an article in
the *Annals of the Association of American Geographers* entitled "Major
Agricultural Regions of the Earth." This is still the basic work for the
map of Major Agricultural Regions which appears in the widely used
Goode's World Atlas. Concurrently with the work of Jones and Whittlesey, O. E. Baker developed a classification of *Agricultural Regions of
North America* which was published in a series of articles appearing
in *Economic Geography* starting in October, 1926. This classification
by Baker formed the framework for the more detailed and refined
classification of types of farming published by the U.S. Department of
Commerce in 1933.

Whittlesey recognized two main forces in his classification: (1) "the
combination of environmental conditions which sets the limits of range
for any crop or domestic animal and provides within these limits, op-
timum habitats" and (2) "the combination of human circumstances
which applies the habitat possibilities of plants and animals to human
needs." Whittlesey identified the "chief elements of the cultural circumstances" as being "density of population, stage of technology, and
inherited tradition." The interaction of these forces produces the "func-
tioning forms which appear to dominate every type of agriculture."
These forms are listed by Whittlesey under five headings:

1. The crop and livestock association
2. The methods used to grow crops and produce the stock
3. The intensity of application to the land of labor, capital, and
   organization, and the outturn of product which results
4. The disposal of the products for consumption (i.e., whether used
   for subsistence on the farm or sold off for cash or other goods)
5. The ensemble of structures used to house and facilitate the farming operations[1]

---

[1]Derwent Whittlesey, "Major Agricultural Regions of the Earth," *Annals of the
Association of American Geographers,* Vol. 26, December, 1936, pp. 208-09.

Baker in his regionalization of American agriculture defined an agricultural region as "a large (sub-continental) area of land characterized by homogeneity of agricultural conditions, especially crops grown, and sufficient dissimilarity from conditions in adjacent territory as to be clearly recognizable." Baker goes on to state that "agricultural regions are determined principally by climatic conditions." Then he emphasizes that

> Within an agricultural region differences in land relief, or in slope, and in soils may cause such variations in the proportion of the land used for crops, pasture, or forest, or in the relative importance of the crops, as to require recognition, and the consequent division of the agricultural region into sub-regions or areas. The physical conditions in these sub-regions, in turn, especially the soil conditions, may induce such differences in utilization of the land and in the system of farming as to require the division of the sub-region into districts; and these districts, in turn, may be subdivided into localities. Indeed, the division in actual practice is often carried down to a single farm, almost every farmer recognizing differences in the quality of his land that require specialization in utilization and management.[2]

Although both Whittlesey and Baker recognized that further refinements could be made in their classifications by using more detailed criteria for identifying and delimiting major agricultural systems or types, they did not pursue this matter further in their original classifications by developing a hierarchy of sub-types and sub-areas. Recently under the sponsorship of the International Geographical Union, a Commission for Agricultural Typology has been at work trying to develop a more refined and more meaningful classification of agriculture at international, national, and local scales.

In order to make a more quantitatively refined classification, many aspects of agriculture must be compared and evaluated from place to place. Some of the conditions that must be considered and assigned a place on a scale of significant values are listed below along with some selected relevant data about these conditions in American agriculture.

1. *Size of the land holding used for an agricultural activity.* In the United States a great range in farm size may be found in different parts of the country and for different kinds of products being produced. Thus in 1964 the average size of farms in North Carolina was 97 acres; New Jersey, 109 acres; Illinois, 225 acres; Nebraska, 596 acres; and Wyoming, 4,100 acres. Poultry and tobacco farms are generally small in size, and 73 percent of both poultry and tobacco farms were

[2]Oliver E. Baker, "Agricultural Regions of North America: The Basis of Classification," *Economic Geography*, Vol. 2, October, 1926, pp. 468-70.

less than 100 acres in size in 1964. On the other hand, 73 percent of all cash-grain farms were more than 100 acres in size.

2. *Tenure of the farm operator.* In 1964, 58 percent of all farms in the United States were operated by full owners, 17 percent by tenants, and nearly all of the rest by part owners. However, only 29 percent of all land in farms was operated by full owners, whereas 48 percent of the farm acreage was operated by part owners, 13 percent by tenants, and 10 percent by managers.

3. *Use of land resources in farms.* In 1964, American farmers used the 1,110 million acres of land in farms as follows: Cropland, 39 percent; pasture (not cropland and not woodland), 44 percent; woodland, 13 percent; farmsteads, farm lanes, drainage and irrigation ditches and other uses, four percent. Great regional contrasts do exist. In the Corn Belt states of Ohio, Indiana, Illinois, Iowa, and Missouri, 72 percent of the land in farms was used as cropland, whereas in the Mountain States of Montana, Idaho, Wyoming, Colorado, New Mexico, Arizona, Utah, and Nevada only 16 percent was so used.

4. *Value of land resources used for agriculture.* In 1964 the Bureau of the Census reported the value of farm real estate, which includes land and buildings at 160 billion dollars. The per acre value of farm land and buildings of course varies greatly from state to state. Thus in Iowa, with highly productive land, the average value was $254. In Montana, where much less of the farm land can be used for crops and at greater risk unless irrigated, the average value was only $35. On the other hand, farmers were using land for agriculture in states such as Florida, where prospects of future use for homes and other non-agricultural uses has created a per acre value of $217.

5. *Use of hired and family labor.* In 1964 during the week preceding the enumeration which was made in autumn, 2.3 million out of 3.2 million farm operators reported working 85 million hours. Persons 14 years old and over in the households of farm operators totaled five million of which 2.2 million worked 41 million hours during the week preceding census enumeration. Hired workers who worked a total of 150 days as farm laborers totaled 890,000 in 1964. California, Texas, and Florida with 122,000, 76,000, and 55,000 regular hired workers respectively led all other states. In these states, fruits and vegetables are important crops in some areas.

6. *Investment of capital in equipment and facilities.* American farmers have greatly expanded their capacity to produce agricultural commodities by making heavy investments in machinery and facilities. In 1820, each farm worker in the United States produced enough farm products for four persons. By 1900, eighty years later, he was producing enough for seven persons, but by 1964 he was supplying 33

persons. A great range of equipment and facility items have contributed to this increased productivity of farm workers. Among some of the more important are: tractors, motor trucks, automobiles, mechanical planting, cultivating, and harvesting equipment, electrically powered facilities, and telephones.

7. *Expenditures for feed, livestock, fertilizers, seed, fuel and oil, machine hire and custom work, and hired labor.* In 1964, American farmers spent 17.6 billion dollars, which amounted to half of the total value of all farm products sold, on these seven major items for the farm business. The breakdown was as follows:

	Millions of dollars
Feed for livestock and poultry	5,512
Purchase of livestock and poultry	4,177
Fertilizer and fertilizing materials	1,772
Seed, bulbs, plants, and trees	661
Gasoline and other petroleum fuel and oil for the farm business	1,787
Machine hire, custom, and contract work	870
Hired farm labor	2,799

8. *Improvements to land.* Agricultural land is improved by irrigation, drainage, use of chemicals, rotation of crops and pasture, and application of such practices as contour planting and stripcropping, construction of ponds and reservoirs, and numerous other improvements that contribute to the increased productivity of labor, land, and capital used in agricultural activities.

9. *Use of credit in producing agricultural commodities.* Two main types of credit are important in American agriculture. Long-term loans are needed by farmers to buy land. These loans need to be arranged in such a way that farmers are able to repay them over sufficiently long periods of time so that there will be no undue strain on their financial resources. Money is also needed to buy equipment and the materials necessary in producing agricultural commodities from year to year. This production credit needs to be readily available at reasonable interest rates if it is to be useful in meeting farmers' needs at critical times. In 1966, the farm mortgage debt in the United States amounted to 21 billion dollars. These were loans that were secured by farm real estate. An additional 19 billion dollars of several different kinds of loans not secured by real estate were outstanding as of January 1, 1966.

Quantitative recognition of these important conditions, which have just been outlined, is much easier in some parts of the world than in other areas where inadequate statistical information still limits greater refinement in the regionalization of agricultural activities. A number of other dimensions also must be recognized.

One of the most important dimensions that has often been neglected in the geographical literature is the role of government policy in determining many aspects of agricultural production. The collectivization of agriculture in the Soviet Union and in countries of eastern Europe is an example of collective decision-making that affects significantly the pattern of agriculture in that part of the world. In the United States many Federal programs exert a major influence on production of many agricultural commodities. The greatly increased importance of soybeans and sorghums in areas where corn, wheat, and cotton are major crops is an example of this influence. The decision to give Federal support to irrigation in western states and to flood control in the Mississippi Valley is yet another example of the role of the Federal Government in affecting the decisions of individual farmers. The collective effect of Government policies can of course be very important in determining the regional composition of agricultural activity.

It is also of great significance in studying the regional characteristics of agriculture to understand and evaluate the relation of the agricultural sector of economic activity to other sectors of the international, national, or local economy. The growing urbanization of life has had a major influence on agricultural activities in the United States. The labor supply for agriculture, the value of land used in producing agricultural commodities, and the location of producing areas in relation to markets are some of the ways in which this influence is felt. The unionization of labor in manufacturing and mining industries has affected the level and distribution of agricultural activities.

Farmers must sell their produce in a predominantly competitive market without really an effective means for bargaining for higher prices because there is no strong organization of farmers to exercise bargaining power. Thus an imbalance often develops between the prices paid by farmers for products produced in other sectors of the economy with strong unionization of labor and a structure of monopolistic competition among many industries and the prices which they receive in selling their products on a competitive market. Some of the Federal farm subsidy programs have as a major objective the attainment of parity between the agricultural and other sectors of the economy.

In classifying or typing agricultural functions and forms, the spatial arrangements and relationships of these characteristics need to be studied more carefully, and greater refinement in measurement will be desirable. For example, the spatial arrangement of broiler production in the United States is quite different from the distribution of cattle production. Furthermore, this pattern of broiler production has been undergoing change over a period of time because additional factors have been introduced. The pattern of soybean production in the United States has still another configuration that differs markedly from that for broiler and cattle production. More refined analysis of these configurations and interrelationships among them offer profitable future research directions for geographers.

A Framework for Classifying World Agriculture

In recent years, a number of introductory textbooks in the field of economic geography have presented the discussion of agriculture on a global scale with categorizations of agricultural systems or types that closely resemble the Whittlesey classification developed more than thirty years ago. The high degree of uniformity that still exists in the typing or classification of agricultural activities serves to confirm the careful scholarship of Professor Whittlesey. The fact that few major adjustments or changes in the basic classification have been needed also is indicative of the stability of agricultural systems, many of which have evolved over long periods of time.

Yet both subtle and major changes in the types or systems of agricultural activity have occurred at all levels from international and national to local. One of the most significant changes that has taken place at the international level is the collectivization of agriculture in the Soviet Union, parts of eastern Europe, Cuba, and eastern Asia. Although the crops and livestock products have not necessarily changed greatly in these parts of the world, there have been major changes in decision-making and organization associated with agricultural activities under a collective scheme of operation.

In the classification of agricultural systems at a world scale, three major levels have been consistently recognized: ecological, subsistence, and commercial. At the ecological or near-ecological level, people are mainly utilizing the natural plants and animals of the area in which they are living under primitive conditions. At the subsistence level, which has generally been subdivided into at least primitive and intensive subsistence sub-types, the farm products are being raised primarily for consumption on the same farm units on which they have been produced. Relatively little sale of farm products even occurs at the community or local levels and practically none of the agricultural commodities being produced at the subsistence level enter into world trade. At the commercial level, agricultural production is primarily for sale, although some farm products are consumed on the same farm units on which they are produced. To these three consistently recognized levels of agricultural systems, the author recognizes the collective agricultural system as being a fourth level that should be added.

Often overlooked in presenting a regional picture of agriculture at a world scale is the intermixture of the ecological, subsistence, commercial, and even collective levels that exist in some parts of the world. On a world map made at a small scale for publication in a book or for use in the classroom, it is not possible to recognize the intermixture of these levels that do exist. The most striking intermixture of levels occurs in tropical parts of the world where plantation commercial agriculture has been introduced into areas having ecological and subsistence systems of agriculture. Rubber and tea plantations in southern and eastern Asia, banana plantations in Middle America, and sugar plantations in the Caribbean are striking examples of such intermixture. On the other hand, examples may also be cited of agricultural systems which are not intermixed to any great extent with systems operating at different levels. The Corn-Soybean-Livestock system of the American Midwest, and the Grain-Livestock system of the Pampa area of Argentina are examples of commercial systems that overwhelmingly dominate the regions in which they occur.

The framework of agricultural systems at a world scale which has just been outlined can be completed in several different ways. The author has a preference for the classification presented below. However, it is clearly recognized that this or other classifications presently in widespread use have certain limitations and that further research is needed for refining and reformulating a more meaningful and more detailed classification of agricultural systems.

a framework of agricultural systems

Ecological or Near-Ecological Primitive Migrating Systems
Gathering (often combined with hunting and fishing)
Nomadic Grazing
Shifting Cultivation (sometimes classed as primitive migratory subsistence)

Subsistence Systems
Primitive Sedentary Subsistence
Intensive Sedentary Subsistence
Near-Subsistence Mediterranean

Commercial Systems
Tropical and Subtropical Plantation
Mid-Latitude Grain Farming
Vegetable and Fruit Culture

Mixed Crop and Livestock Farming
Dairy Farming
Livestock Ranching

Collective Systems

These systems parallel several of the commercial systems in terms of the crop and livestock products being produced but differ markedly in organization and decision-making in the production process.

World Agricultural Systems

Space does not permit a detailed description of the character and distribution of each of the complex systems which has just been outlined. A detailed analysis of these systems would of course be complementary to the product by product analysis of world agriculture of Chapter 3. However, a descriptive outline is provided in order to guide further study of the agricultural systems under which the many agricultural commodities are being produced.

ecological or near-ecological systems

In general, the agricultural systems existing at this level are characterized by (1) provision of only the most rudimentary human wants, (2) location in environments that are not much used at higher levels of agricultural production, (3) following custom and tradition is more common than accepting change, (4) physical or cultural isolation from the rest of the world, (5) much reliance on untended products of the natural environment, and (6) low man-land ratio or sparse population in relation to area occupied.

Gathering

Gathering, which is often combined with hunting and fishing is found mainly in the (1) high northern latitudes, (2) tropical Amazon and Orinoco basins of South America, (3) Congo Basin, Guinea Coast, Kalahari Desert, interior Mozambique and adjacent areas of Rhodesia and Malawi, (4) interior parts of southeastern Asia, (5) interior areas of Indonesia, (6) New Guinea, (7) interior western Australia and the Arnhem Land area of northern Australia.

In the northern high latitudes the Eskimos depend mainly on hunting and fishing but in some areas people such as the Lapps have herded reindeer for a living. In isolated parts of the wet tropics, fruits, nuts, and other edible parts of plants along with animals and fish are the main sources of food. Some of the Pygmies of Africa and the Semang of Malaya and Thailand are examples of inhabitants in the wet tropics living this kind of life. In dry tropical and subtropical areas the Bush-men of the Kalahari and the aborigines of Australia live a nomadic type of life. At certain times the Bushmen get food from big game animals but at other times of the year they depend upon such things as ant eggs, succulent roots, worms, frogs, lizards, and snakes for food.

Nomadic Grazing

Nomadic grazing is concentrated mainly in Saharan Africa, east-ern Africa, interior steppes of southern Africa, southwestern Asia, and interior Asia including the central Asia area of the Soviet Union. In order to get sufficient forage and water for their animals, it is neces-sary for the nomad to keep moving. The camel is used extensively in some areas. Goats, sheep, and cattle (in areas where there is more moisture) are animals used in most nomadic grazing areas and milk and wool or hair and even blood are animal products utilized, as well as their meat, by nomadic peoples. Some nomadic groups prefer to rely on one animal, such as the Bedouins with their camels and the Masai, Herero, Hottentot, and other groups of southern and eastern Africa with cattle; but most nomadic groups herd different kinds of animals at the same time as they move from place to place.

Shifting Cultivation

According to the Food and Agriculture Organization of the United Nations, about 200 million people are making a living by practicing some kind of shifting cultivation. The shifting cultivator basically car-ries out a long-term rotation of crops with the natural forest vege-tation. Patches in the forest are cleared and burned and crops are planted. After one to five years, these crop plots are permitted to revert to forest to be cleared again from five to twenty-five years later de-pending on population density and the rate of fertility renewal. The shifting cultivator raises crops almost entirely for his own use. In some places the shifting cultivator moves both his house and the plots of land being used for crops. In other areas, only the plots used for crops are shifted. Shifting cultivation is a type of agriculture found over ex-tensive areas of the wet tropics of South and Middle America, Africa, and southeastern Asia. Crops grown, tools used, and farming practices differ from one area of shifting agriculture to another.

Professor Crist has carefully described the migratory or shifting culti-vator's use of his small plot or plots of land, which generally has been laboriously cleared by hand and then burned. His description is applic-able to extensive areas of the wet tropics of Latin America.

> Over millennia this primitive farmer has evolved a sure-fire crop com-plex: Corn, yuca, beans, and pumpkins or squash. To this list has been

added, since the Spanish Conquest, the cooking banana. The techniques of this kind of primitive farming look simpler than they are. The great toe of the farmer's bare foot is used—or a dibble stick if one is fancy—to make a hole in the soft earth of the burnt-over plot, preferably after the first rain; two or three grains of corn and a few beans and squash or pumpkin seeds are put in, and the hole is covered up by a swipe of the foot and the earth is tamped down by being stepped on. The corn comes up rapidly and shoots up fast, its green stalk forming a living pole for the beans to climb up on; it will be harvested in about 90 days, the beans a month or so later. The squash or pumpkins are ideal for this kind of farming, for their vines spread all over the ground and over the fallen logs and heaps of brush that have not been burned. Thus every bit of land is actually in use, and each crop gets its quota of sunshine. Near the house, especially where the slops are thrown, will usually be the patch of cooking bananas, one of the first plants to be planted and a great producer of food for all. One of the most significant food plants is yuca, a patch of which is usually found near most huts.

The subsistence farmer is in a closed, almost hermetically sealed, economic unit or cocoon that he spins around himself; he may try to work his way out of this cell by taking to market a bag of raw cotton, or perhaps an extra bunch of cooking bananas, or even a few kilos of yuca, but these small surpluses usually are not produced by design. The load of firewood or charcoal, however, carried by mule or canoe or even on one's back, is expressly cut or made for sale and is often the only and seemingly very tenuous economic thread that gives the slash-and-burn cultivator contact with his fellows on the regional scene.[3]

subsistence systems

The subsistence agriculture systems have several characteristics in common. All or nearly all of the products grown on subsistence farms are consumed on the farm by the operator and his family. The farms are small and may range from as little as two or three acres in size in areas of intensive subsistence agriculture to 15 to 25 acres or more in areas where more primitive conditions prevail, where population density is not so great, or where the natural environment makes larger farms necessary. Draft animals are used in some subsistence agriculture but relatively little in other areas. High inputs of manual labor relative to the use of capital for labor-saving machinery is a dominant characteristic. Equipment used in farming is primitive by comparison with the machinery used on farms in the United States. Hoes, spades, walking plows (sometimes made of wood), and the hand sickle are in common use. Animals, which are more common in highland areas of subsistence agriculture, are of poor quality and are few in number per farm. Cereal and root crops are the basic crops grown. In the Medi-

---

[3]Raymond E. Crist, "Tropical Subsistence Agriculture in Latin America: Some Neglected Aspects and Implications," *Smithsonian Report for 1963*, Washington, D. C.: Smithsonian Institution, 1964, pp. 505-06.

terranean area the olive is an important crop and in the Andean highlands potatoes are much used.

Primitive Sedentary Subsistence Agriculture

A rather fine line separates the primitive shifting cultivation system
of agriculture and the primitive sedentary subsistence type where the
operator remains in place on the land. Primitive sedentary agriculture
is found most commonly in the highland areas of the tropics; however,
some farmers of the wet tropical lowland areas have acquired enough
understanding of the rudimentary principles of soil science to enable
them to remain in one place. Increases in the population density make
it essential for people to stay on a particular tract of land on which
they at least can exercise squatter's rights. In some places along streams
or near lakes, farmers are able to supplement their food supply by
staying in one place. Mineral exploitation by foreign enterprise and
the opportunity to sell a few items such as spices, rubber, and cinchona
have also been factors in determining where the sedentary approach
to farming prevails over the migratory type of agriculture. Development of new transportation routes into previously isolated areas has
also offered a strong incentive for farmers to "settle down."

In the highlands and plateaus of the tropics, elevation provides an
amelioration of temperature and humidity conditions found in the wet
lowlands. It is in these plateau and highland areas of Africa, southeastern Asia, and the Americas that primitive sedentary agriculture is
most important. In some places where plantations growing commercial
crops are nearby, farmers have broken the complete reliance on their
own small farms for a livelihood by working also as laborers on the
plantations.

Intensive Sedentary Subsistence Agriculture

Between a third and a half of the world's people and about two-
thirds of the world's farmers are living on intensively operated subsistence farms. Most of these farms are located in southern and eastern
Asia. The Tigris and Euphrates valleys of Iraq and the Nile Valley
of Egypt are the only other areas where this type of agriculture is of
widespread importance. In these parts of the world the high man-land
ratios make an intensive use of land resources an economic necessity.
Over the centuries a very careful tillage of the soil and the improvement of land resources has increased the area of arable land and the
usefulness of this land for agricultural production. In some parts of
eastern Asia, steeply sloping land has even been diligently terraced
with great expenditures of human labor over the years in order that
effective use of such land can be made for crop agriculture. In the

world's main areas of intensive subsistence agriculture the use of irrigation is a widespread although not a universal practice. Again, water is often applied to the fields by the use of much manual labor.

Characteristic of most of the areas of intensive subsistence agriculture is a village pattern of settlement. The farmer and his family live in a small village with other farmers and their families. Each day the laboring members of the family, which includes women and children, travel back and forth to the fields where their crops are planted when there is work to be done. The fields that belong to a single family often are scattered, and noncontiguous tracts have been decimated and reduced in size by long-standing inheritance practices associated with passing on the highly respected and treasured family-owned land-holding from one generation to the next. Farms are now very small, many of them being no more than three to five acres in size. Some are even smaller, although in other areas the average size may be larger.

In those parts of the world where intensive subsistence agriculture is dominant, crops furnish the main source of food. Grains are the principal crops grown, and in southern and eastern Asia rice is the most widely produced of all the grains. In some areas such as northern China and in northwestern India and adjacent western Pakistan, wheat is a very important crop. Barley, sorghums, and millet are also grown for human food where rice cannot be grown. Other crops such as the pulses, yams, sugar cane, corn, fruits, and vegetables are also grown. Cotton is an important fiber crop in some areas such as the Nile Valley and in India and in thees areas furnishes a source of cash income that helps break the complete reliance on products raised on the farm.

Intensive crop cultivation practiced in areas of dense population requires a large input of human labor, which generally has only primitive implements available, in order to obtain as much food from the land as possible. Some use of cattle and water buffalo as well as other animals is made for draft purposes in some areas. Literally hundreds of man hours are spent in the production of crops on the small farms that are the sole source of food for millions. Very little use is made of livestock for food, since the amount of food available from meat, milk, eggs, and other livestock products is far less than that obtained through the direct human consumption of the crops grown. However, a few animals per farm, such as a hog or two, some poultry, or some goats are often used as scavengers to make use of crop and other vegetative wastes. In areas such as India, cattle are kept in large numbers, some of which are grazed on non-arable land.

In addition to the intensive use of labor and such practices as irrigation of cropland and the terracing of steep slopes to make them arable, the use of manure is an important key to the maintenance of the productive capacity of the land under continuous crop culture. Green manures obtained from various kinds of plants, animal manures, and night soil (human excreta) used in some areas are carefully applied to obtain the best yields possible. Still other practices that are often used are multiple cropping and intertilled cropping. Multiple cropping is the practice of raising more than one crop on the same plot of land during the same year. This practice is possible in subtropical and tropical areas where adequate moisture is available either naturally or by irrigation, since there is no frost hazard. Intertillage is the practice of planting two or more crops in the same field at the same time.

When taken together, these many practices and the great application of human effort amount to a remarkable and really quite complex system of agriculture which makes it possible for millions of people to subsist on land resources which appear to those in the World of Plenty to be woefully inadequate to support so many people having only the primitive implements and traditional practices, which are characteristic of this part of the World of Poverty.

Japan is an interesting exception to the intensive subsistence agricultural patterns described above. Actually the Japanese farmer is no longer the subsistence farmer that he was 25 to 40 years ago. With the rapid increase of small tractors and other innovations on Japanese farms, farming in that country has rapidly broken out of the traditional subsistence pattern to become a semi-commercial or semi-subsistence type. Many of the characteristics of subsistence agriculture still remain such as the relatively small farm unit and the dominance of crops, but the sale of a considerable amount of farm produce now is characteristic of many Japanese farms.

Near-subsistence Mediterranean Agriculture

In parts of the Mediterranean world of southern Europe, western Asia, and northern Africa a type of agriculture has developed over hundreds of years that is distinctive from the agricultural types found anywhere else. The dry summer subtropical climate in large measure explains this particular pattern of agriculture. Also responsible for this distinctive type is the long history and the role that tradition has played. In other areas of the dry summer subtropical climate, which has a winter concentration of the 15 to 25 inches of rainfall that generally can be expected annually, similar crops are grown and some of the agricultural practices followed do bear some resemblance to those used in the Mediterranean area. However, in southern California and the Central Vale of Chile, there are great contrasts with the agriculture of the Mediterranean. In California a high degree of commercialization is associated with the production of fruits, nuts, and vegetables in a sub-tropical climate for the very large urban markets of mid-latitude areas of the United States. Most, but not all, of the people who live in these urban centers can afford to purchase fresh vegetables grown out of season, fresh subtropical fruits such as oranges and grapefruit, and processed fruits and vegetables grown in these areas.

In sharp contrast to the Mediterranean commercial agriculture of southern and central California, the agriculture of the island of Cyprus in the eastern Mediterranean is described as follows in a recent report of the Foreign Regional Analysis Division of the Economic Research Service of the United States Department of Agriculture.

> Cypriot agriculture is largely on a family-enterprise basis. Approximately 85 percent of all farmland is worked by owners; about 10 percent is leased (half under long-term leases); the rest is sharecropped. In a part of the world where fragmentation of farmland is more often the rule than the exception, the Cypriot system of landholding offers an extreme example of subdivision compounding subdivision.
>
> With expanding population and static land area, most of the island's farms have shrunk through the years to a very small size. Only one-fifth of the cultivated acreage lies in farms of more than 70 acres; more than half the farms are smaller than 12 acres; about 15 percent have 3 acres or fewer.
>
> And because traditional laws of inheritance, under which all children and their heirs share equally, the average holding may be divided into more than a dozen plots—some scarcely large enough to accommodate oxen and plow—belonging to as many different farm families. In addition, separate plots of land owned by one farmer may be miles apart. The patch-work composition of the majority of Cypriot farms brings inherent inefficiencies in time and effort, makes it difficult to understand and employ water rights, and severely limits use—or the potential for use—of farm machinery for expansion of output.
>
> Cyprus' main crops, wheat and barley, are grown under winter rainfall, sometimes with supplementary irrigation. A wide range of pulses for both food and feed are also grown in winter, but on a much smaller scale than grains. Spring-sown crops, started with late rains or on spate-irrigated land and then grown under dry-farming conditions, are mainly cotton, cuminseed, chickpeas, sesame seed, and tobacco. Summer crops grown under full irrigation include cotton, cowpeas, haricot beans, peanuts, melons, strawberries, potatoes, carrots, and a variety of other vegetables. These and other permanent crops are cultivated principally in the plains. Almonds and olives, usually produced without irrigation, are widely distributed. But citrus is found chiefly in coastal areas with a mild climate, where irrigation water is available in sufficient quantities for presently bearing groves.
>
> The carob tree (*Ceratonia siliqua*) thrives on the seaward slopes of the two mountain ranges up to altitudes of about 1,000 feet; its beans are used for feed. The valleys on the southern massif are intensively cultivated; vineyards cover many slopes. Grapes are produced on an area greater than that of any other crop except grains. Cherries, apricots, plums,

peaches, apples, and other deciduous fruit trees are also grown under irrigation, mainly in the valleys of the southern hills.

Free-ranging, fat-tailed sheep and small, hardy goats make up the bulk of the island's livestock. Sheep-raising predominates in the lowlands; goat-raising tends to be concentrated in the higher, more inaccessible areas of the country. Both sheep and goats are very largely maintained on waste-land, weed growth, and crop residue grazing.[4]

With local variations this is the type of agriculture that still is present over much of the Mediterranean area. While this agriculture is less fully subsistence in character than the intensive subsistence types previously described, it is much more closely aligned with subsistence systems than with the commercial types to be discussed next.

commercial systems

In contrast to the ecological and subsistence agricultural systems just described, the commercial agriculture of the world is characterized by the production of agricultural commodities mainly for sale. In commercial agriculture, much use of capital is made for the purchase of tractors and machinery, fertilizers, insecticides, pesticides, herbicides, improved plants, better breeds of animals, and many other technological innovations. In looking at commercial agriculture at a world scale, there are many variations in the application of technology to farm operations that must be recognized. The commercial farmer is much more dependent upon the world around him because he not only seeks a market for his products but he must buy many things for the farm operation and for the personal use of his family.

The principal systems of commercial agriculture are only outlined briefly here, since much of Chapters 2 and 3 deal with the commercial production of agricultural commodities.

Tropical and Subtropical Plantations

These units specialize in a number of different products. Historically, plantations have employed relatively large numbers of unskilled workers to produce tropical and subtropical crops for export to mid-latitude areas. Ownership of these plantations and the supply of capital needed for their operation generally has not been indigenous to the areas where the plantations are located. An exception to these latter conditions was the cotton plantations of the southern United States. The main areas having the several main specialities are:

1. Natural rubber plantations of Indonesia, Malaya, Thailand, Ceylon, Vietnam, the island of Borneo, Cambodia, India, and Burma.

---

[4]H. H. Tegeler, "Cyprus Agricultural Economy in Brief," *Foreign Agriculture Economics,* Washington, D. C.: Foreign Regional Analysis Division, Economic Research Service, United States Department of Agriculture, July, 1966, pp. 4-6.

Rubber is also produced on small holdings in some of these countries.

2. Banana plantations located mainly in coastal areas in Middle America, Ecuador, Brazil, Colombia, Caribbean Islands, West Africa, Mozambique, Somaliland, and Taiwan. As in the case of rubber, bananas are also produced on small farms in these areas.

3. Tea plantations are located principally in Java and Sumatra in Indonesia; Assam, West Bengal, Madras, and Travancore-Cochin areas of India; and Ceylon.

4. Sugar plantations have been particularly important in Cuba, Puerto Rico, and other Caribbean areas, Brazil, Hawaii, and the Philippines.

5. Cotton plantations of the southern United States. Historically these plantations were more important than at present; however, particularly in the Lower Mississippi Alluvial Valley, large plantations are still major producers of cotton.

Mid-latitude Grain Farming

This type of agriculture is characterized by heavy reliance on wheat and other small grains, which are produced on highly mechanized farms mainly in areas having a relatively sparse population. The semiarid and subhumid areas of the mid-latitudes in the United States, Canada, Australia, and Argentina are the major areas. Under the collective system of agriculture, large-scale grain production is found in the famous chernozem (black earth) and related soil regions of the Soviet Union. In Anglo America, sorghums are associated with the production of wheat in the southern Great Plains and with barley in the northern Plains.

As a boy the author worked on his uncle's wheat farm south of Saskatoon in Saskatchewan. At that time (1936), the farm was comprised of five sections (about 3,200 acres) of land with highly fertile chernozem soils. Well-remembered is the unreliable rainfall condition of the 1930's which greatly troubled large-scale grain farmers of the Great Plains. Good yields on the farm in Saskatchewan during the 1920's were followed by several years of complete or nearly complete failure of crops during the early part of the 1930's. In 1936 the yields were good and the price of wheat was also good, so this uncle was able to repay some of his debts incurred during previous bad years. That year the author's uncle had planted about 1200 acres of wheat, 200 of which were durum wheat; 300 acres of barley; and 200 acres of oats. About 1300 acres of land was in summer fallow that year. In 1936 wheat was still being bundled, shocked, and later threshed by a stationary threshing machine. Today on the same farm, large moving combines cut and thresh the wheat in one operation. A similar approach to the production of wheat and other grains is found in other semiarid and subhumid parts of the world.

Vegetable and Fruit Culture

Vegetables and fruits are produced and sold from farms of many different sizes. The scale of operation varies from such large corporate or family-owned units as the pineapple plantations of Hawaii and the thousands of acres of citrus in Florida under one management to the small family-operated market gardens found close to many European cities. A great variety of fruits and vegetables are produced and sold in many different ways and from markedly different circumstances of location. Fruits are sold fresh, frozen, dried, or canned. Exotic tropical fruits are now flown by air to markets in large cities in the mid-latitudes. Large quantities of vegetables are shipped fresh under refrigeration from Florida, Texas, and California during the fall, winter, and spring months to northern markets. Quick-freezing and canning of vegetables are also of major importance in the United States but are considerably less significant in many other parts of the world. Some fruits and vegetables are produced under highly artificial conditions such as in greenhouses and with the use of plastic and cheesecloth coverings. In other instances the production is mainly concentrated in those areas having the most favorable natural conditions for efficient production.

Mixed Crop and Livestock Farming

On a commercial basis the mixing of crop and livestock production on the same farm unit is most readily observed in the eastern United States; western, central, and eastern Europe; South Africa; and in southern Brazil, northeastern Argentina, and in Chile south of about the 38° parallel. Generally the farms in areas of mixed crop and livestock farming are not as large as in the areas of commercial grain farming. The degree of mixture varies considerably from place to place. In the American Midwest, corn, soybeans, alfalfa along with some oats and wheat are the main crops, and hogs and beef and dairy cattle are the main livestock. However, any one farm is not likely to have all of these different crops and animals. Corn, soybeans, and hogs may be the main products of one farm while on another farm corn, oats, alfalfa, and beef cattle may be dominant. Generally, oats and wheat are likely to be more important when alfalfa is being grown on farms where beef or dairy cattle are raised. Formerly, poultry and egg production was a part of the mixed farming pattern of the United States, but now these products are being produced mainly in large volume on highly specialized units. Sub areas of different combinations of crops and livestock may be observed in the Midwest.

In Europe where units of mixed crop and livestock farming are smaller than in the United States, there is less emphasis on cattle and hogs. However, many European farmers like to keep some animals on their farms in order to have manure available for fertilizing their crops. Sugar beets, potatoes, and rye are important crops on many farms. European areas of mixed farming are also characterized by less mechanization than American farmers.

Much change has taken place on farms of the American Midwest during the past quarter of a century. When the author was a boy growing up on a mixed crop and livestock farm in central Indiana, his chores and summer work assignments at the age of fourteen included: Helping milk five or six cows twice a day, feed 40 to 60 head of hogs, feed 10 to 15 steers, feed and care for 8 to 10 head of horses which were used as draft animals, feed 150 to 200 chickens, gather eggs from 100 to 125 laying hens, help pick tomatoes after school in the fall of the year, plow corn in June, shock wheat bundles in June, haul hay shocks during hay-making in July and August, help kill hogs and a steer for meat in winter, help can fruit and vegetables (tomatoes, beans, apples, peaches, blackberries), and churn butter.

Today the farmer who operates the same farm of 160 acres on which the author was reared plus an additional 200 acres has a far smaller variety of work for his son to do. There are no horses to feed, no cows to milk, no chickens to feed, no eggs to gather, no tomatoes to pick, no wheat to help harvest, no vegetables or fruit to can, no butter to make, no hogs or steers to kill. The present operator of the farm plants about 165 acres of corn and 100 acres of soybeans, raises about 40 acres of hay, and has about 50 acres of pasture. The only livestock kept are hogs and steers. The farmer's wife buys her meat, butter, fruit, vegetables, and eggs at the supermarket in the county seat. Thus, today there is a much higher degree of specialization on most farms which produce a mixture of crops and livestock in the Midwest.

Dairy Farming

There are three major regions where commercial dairying is an important type of farming: (1) northwestern Europe, southern Scandinavia, and the Baltic-Moscow area of the Soviet Union, and the Baltic area of East Germany and Poland; (2) the northeastern states and the southern parts of the Lake States (Wisconsin, Michigan, and Minnesota) in the United States; (3) New Zealand, Tasmania, and adjacent coastal areas of Australia. In the United States, secondary smaller areas of considerable importance are located in California and in the Puget Sound-Willamette Valley areas of Washington and Oregon. Of all the major dairy regions of the world, New Zealand is unique in having ideal physical conditions permitting year-long grazing but it has the extreme disadvantage of producing a highly perishable commodity at great distance from urban markets. Thus most of New Zealand's milk is made into butter. In the United States, dairying has been particularly important north of the line marking 130-140 frost-free days a year, which is about the lower limit for producing corn for grain. Significant is the fact that the States of the Lower Great Lakes, Middle Atlantic, and New England regions have many of the nation's great industrial and commercial centers.

Although milk can now be dried, powdered, evaporated, and otherwise processed into a non-perishable form for marketing, the sale of fresh fluid milk is still the most common way of marketing the main product of dairy farms. Because of the high costs associated with marketing fresh fluid milk, milksheds around major urban centers may easily be observed in the United States. The size and shape of these milksheds will vary with the size of the city being served. Generally a circular or nearly circular shape for a milkshed will prevail unless natural conditions are unfavorable in a sector of the shed and unless competition from an adjacent city exists. Milk from more distant farms may be sold fresh when supplies of milk from the farms nearer to the market are inadequate to meet the demand. Otherwise, milk may be manufactured into butter, cheese, ice cream, and reconstituted into a non perishable form such as evaporated or dried milk.

Dairying takes much labor and capital. Dairy farmers must work seven days a week all year long or must have someone else milk their cows for them if they want a vacation. Of all the major types of farming, dairying has the advantage of having a more even use of labor throughout the whole year. In commercial grain farming, the farmer is very busy for only about a month at planting time and another month at harvest time. Peak periods when labor requirements are high do exist on dairy farms, such as at hay harvest or silage-making periods; but the farmer who milks cows gets a milk check regularly throughout the year in contrast to the grain farmer who probably will sell his wheat all at one time, since he doesn't even have any storage facilities.

In recent years the need for capital in dairying has increased substantially in the United States. Public health regulations pertaining to the sale of fresh milk are quite strict and must be met by the dairy farmer if he is to sell his milk. Thus, if he is to keep the bacteria count of his milk below the specified allowable level, he will have to invest much money in refrigeration and other equipment which will permit rapid sanitary handling of an easily contaminated product. Another major investment is for large barns and silos for storage of feed and housing of animals during cold winters. The dairy farmer also has to invest in machinery, fertilizers, pesticides, herbicides, and other items needed for the production of crops. In contrast, in New Zealand no barns or storage facilities are needed and relatively little crop production is necessary because of the mild climate.

Livestock Ranching

The grazing of livestock for sale is a type of agriculture that occupies extensive areas in the semiarid and arid regions of the midlatitudes and the savanna areas of the tropics. The main regions are: (1) western United States and Canada and northern Mexico; (2) much of Australia except eastern, southern, and northern coastal areas and the South Island of New Zealand; (3) southern interior Africa south of approximately 12° S. latitude; (4) Llanos of Venezuela and eastern Colombia; (5) eastern Brazil and Bolivia, Paraguay, Uruguay, central Chile, and Argentina except the Pampa; (6) U.S.S.R. immediately east and southeast of the Caspian Sea and the Kirghiz Steppe area.

Livestock ranching is dominant over crop agriculture in semiarid and arid regions which have sufficient moisture for grasses and shrubs that can at least be grazed at certain seasons of the year but which have insufficient moisture for crops, except for relatively small areas where water for irrigation is available. In western United States, extensive natural grasslands at higher elevations provide grazing during the warmer months of the year while the lower and generally drier areas of natural grass and shrubs are grazed at other seasons. Under such conditions the carrying capacity of the grazing land is low and several acres are needed to support an animal unit (equivalent of grazing a full-grown cow for one year). Since many ranches in the American West are unable to raise crops (except where hay may be grown on irrigated land for winter feed), cattle are generally shipped to areas of surplus feed grains in the Midwest or to California to be fattened. Cattle raised on ranches located in the vicinity of larger irrigated areas may also be fattened in feed lots with sugar beet tops, pulp, alfalfa, and some feed grains produced on irrigated land. Some ranches also grow some crops for sale, such as sugar beets for example.

Information taken from a brochure advertising a large ranch for sale in a western state gives some useful insights into the business of commercial livestock ranching. Excerpts from this brochure are quoted here:

> Complete, self-contained cattle operation. Includes approximately 97,000 acres, divided into a summer range of 69,000 acres, a Headquarters Ranch and spring and fall range of about 16,000 acres with a Feedlot on 82 acres, and a winter range of approximately 12,000 acres.
> The ranch has a capacity of 5000 rated units (unit = 1 cow for 1 year). At present, the stock includes a herd of approximately 3000 cows,

developed with great care, and about 185 horses. The ranch produced some 2600 calves in 1958, of which, under the current system of operation 2000 were sold as yearlings at the annual spring auction.

The ranch has an exceptional supply of water from streams and numerous springs, hot springs, and wells. Approximately 3500 acres are now under ditch irrigation.

There are some 30,000 acres of timber on the ranch worth in excess of $1,000,000. In 1959 for the first time, a logging contract is in effect which will produce a regular annual income.

The elevation ranges from approximately 4200 feet to approximately 8400 feet, with an average elevation of about 5500 feet. Average annual precipitation varies from an 8 inch minimum at the Headquarters Ranch to 20 inches to 24 inches on the summer and winter ranges.

The ranch has an abundant free water supply. Its rights on the . . . pre-date all other known rights in the area, and there has never been a shortage or a dispute.

The present manager has an MA degree from one of the leading universities in the Midwest, and he was at one time a professor . . . The availability of labor is good. The present ranch personnel are not tenant farmers. The ranch is now manned by a staff of experts, with practically all key positions held by college graduates.

There are approximately 100 miles of perimeter fencing and 700-800 miles of cross fencing (all 4- or 5-strand wire with steel posts), in good to excellent condition.

The cow herd consists of slightly more than 3000 head of Herefords, including 2600 or more commercial cows and 400 or more registered cows.

There are approximately 150 commercial bulls and 50 registered bulls.

There are tax advantages inherent in a ranch operation. Further information is available upon request.

The ranch is offered as a package at $4,200,000. This includes the cattle and horses, as well as virtually all of the mineral rights. The package also includes the . . . brand of the breeding herd and the . . . brand for the sale stock.

Some of the main characteristics of the different kinds of commercial agriculture in the United States are shown by TABLE 4.1. In this table, selected data for a typical single farm of each of the main types or systems of commercial agriculture are shown. These selected examples must not be interpreted as being typical of the several commercial systems of agriculture found in other parts of the world. In general, it must be clearly recognized that, in the United States, farms have been increasing in size very rapidly. Even more significant is the fact that fewer and fewer farms are now producing more and more of the total agricultural output of the United States. In 1964 farms which sold more than $20,000 worth of products per farm accounted for nearly two-thirds of all farm products sold. Yet these farms comprised only 19 percent of all farms.

TABLE 4.1
Farm Costs and Returns for Selected Commercial Farm Types, United States, 1967

	Unit	Mississippi Delta Cotton Plantation	Pacific Northwest Wheat–Fallow Farm	Corn Belt Hog–Beef Fattening Farm	Eastern Wisconsin Dairy Farm	Northern Rocky Mountain Cattle Ranch
Land in Farm or ranch	acres	1,000	1,550	285	192	5,900
Cropland harvested	acres	609	582	215	135	600
Summer fallow	acres	N.A.	533	N.A.	N.A.	N.A.
Open pasture	acres	N.A.	N.A.	N.A.	29	N.A.
Grazing land owned	acres	N.A.	N.A.	N.A.	N.A.	3,400
Grazing land rented	acres	N.A.	N.A.	N.A.	N.A.	1,900
Crops harvested						
Corn for grain	acres	7	N.A.	87	30	N.A.
Corn for silage	acres	N.A.	N.A.	14	23	N.A.
Wheat	acres	63	511	N.A.	N.A.	N.A.
Oats, Barley, or other small grains	acres	9	49	55	26	19
Soybeans	acres	364	N.A.	29	N.A.	N.A.
Cotton	acres	151	N.A.	N.A.	N.A.	N.A.
All hay	acres	15	22	3	49	336
Livestock on farm or ranch Jan. 1						
All cattle	number	82	38	100	54	406
Hogs	number	18	13	285	29	N.A.
Tractors on farm or ranch	number	6.7	2.1	2.7	2.7	3.0
Total labor used	hour	16,250	4,130	4,300	4,410	6,600
Hired	hour	13,250	540	1,100	440	3,500
Total farm capital, Jan. 1	dollar	382,430	199,400	166,640	89,520	292,690
Land and buildings	dollar	317,000	170,500	116,850	51,640	201,010
Machinery and equipment	dollar	54,160	21,570	18,150	15,390	17,030
Livestock	dollar	9,230	5,900	21,610	14,980	64,900
Crops	dollar	2,040	1,430	15,030	7,510	9,750
Gross farm or ranch income	dollar	78,601	36,452	46,128	22,526	41,438
Total operating expenses*	dollar	40,840	13,064	34,479	12,412	22,373
Net farm or ranch income	dollar	37,761	23,388	11,649	10,114	19,065
Charge for capital†	dollar	27,074	13,763	11,358	5,819	19,757
Return to operator and family labor	per hour	3.56	2.68	0.09	1.08	−0.22

N.A.–Not available or not applicable.
*Includes expenditures for feed purchased, livestock expenses, fertilizer and lime, pesticides and other chemicals, ginning machinery, machine work hired, farm buildings and fences, hired labor, taxes, inventory adjustment, and other.
†At current interest rates.

Source: *Farm Costs and Returns: Commercial Farms by Type, Size, and Location,* Agriculture Information Bulletin No. 230, U.S. Department of Agriculture, Economic Research Service, 1968.

collective systems

In the U.S.S.R., the countries of eastern Europe, Mainland China, North Korea, and North Vietnam, the reorganization of agricultural production under a system of collective and state farms has now been progressing for a number of years. In the Soviet Union the collectivization of agriculture precedes World War II while in the other countries the socialization process is a post-World War II occurrence. Since the collective system has been more fully developed over a longer period of time in the Soviet Union, it is described here as an arch-type of collective agriculture.

In July, 1963, an exchange delegation from the United States Department of Agriculture visited the Soviet Union while a similar delegation from that country was visiting the United States. Excerpts from the U.S. delegation report offer a concise analysis of the collective system of agriculture as the exchange delegation saw it in operation at that time. Selected passages from their report are quoted below:

> Farms in the Soviet Union are huge, both in terms of land area and of the labor force used. Furthermore, all land is nationalized, and agriculture has been collectivized, with the exception of small household plots. There are three types of farm units in the Soviet Union: collective farms (kolkhoz, singular; kolkhozy, plural), state farms (sovkhoz, singular; sovkhozy, plural), and the small private plots permitted members of collective farms, state farm employees, and certain other categories of workers.
>
> The collective farm is the dominant type of farm enterprise from the standpoint of number of units, share of the agricultural labor force, area sown, and output. After fulfilling its obligation to sell stipulated quantities at fixed prices to the state, and of providing for its seed and feed requirements, any remaining production can be sold on private markets. From the total income received, production expenses must be met and approximately one-quarter to one-third of the monetary income is set aside for investment. The remainder of the income is used to compensate members of the farm for their participation in the work. The peasants on collective farms, both male and female, work in the fields under the direction of managers and supervisors, just as workers do in Soviet factories. Payments to the workers vary with the skill required and the amount of labor accomplished.[5]

Over the past few years there has been a merging of many collective farms to form larger units and there has also been a conversion of some collectives into state farms. In 1940 there were about 240,000 collective units but by 1963 the collectives numbered only 40,600 units. Acreage has also declined considerably. In 1953 the sown area was

---

[5]*Soviet Agriculture Today: Report of 1963 Agriculture Exchange Delegation,* Foreign Agricultural Economic Report 13, U.S. Department of Agriculture, 1963.

326 million acres but only 273 million in 1961. Most of this decline in the sown area has occurred since 1958. During the period between 1953 and 1961 the sown area in state farms increased from 45 million acres to 216 million.

REFERENCES

BROWN, LESTER R. *An Economic Analysis of Far Eastern Agriculture,* Foreign Agricultural Economic Report No. 2, United States Department of Agriculture, Washington, D. C., 1961.

CLARK, COLIN and HASWELL, M. R. *The Economics of Subsistence Agriculture,* 2nd edition, New York: The Macmillan Company, 1967.

Farm Costs and Returns, Agriculture Information Bulletin No. 230, Economic Research Service, U.S. Department of Agriculture, Washington, D. C., 1968.

HARDIN, CHARLES M. *The Politics of Agriculture,* Glencoe, Illinois: The Free Press, 1952.

HIGHSMITH, RICHARD M., JR. (editor), *Case Studies in World Geography: Occupance and Economy Types,* Englewood Cliffs, New Jersey: Prentice-Hall, Inc., 1961.

ROEPKE, HOWARD G. *Readings in Economic Geography,* New York: John Wiley and Sons, Inc., 1967.

THOMAN, RICHARD S. and PATTON, DONALD J. *Focus on Geographic Activity: A Collection of Original Studies,* New York: McGraw-Hill Book Company, 1964.

Index

Alfalfa, 41
Arable land, 9, 21, 29
Argentina, 1, 22, 41, 65, 72, 74, 77, 95, 99
Arkansas, 53, 54
Australia, 1, 10, 19, 21, 22, 25, 27, 29, 74, 77, 95, 97, 99

Baker, O. E., 16, 80, 81
Barley, 71
Bolivia, 19, 99
Boulding, Kenneth, 15
Brazil, 19, 22, 65, 67, 72, 74, 95

California, 17, 18, 34, 65, 67, 70
Canada, 1, 9, 19, 22, 59, 68, 95, 99
Cattle, 34, 44-45, 72
Ceylon, 67
Chile, 1, 96, 99
China, mainland, 1, 4, 9, 20, 22, 41, 52, 59, 65, 67, 74
Coffee, 67-68
Collective agriculture, 102-03
Colombia, 19, 67, 99
Corn, 34, 44, 55-58
Cotton, 63-65
Credit, agricultural, 83
Cropland, 9, 21, 22-24, 25, 29
Cuba, 22, 67
Cyprus, 93-94
Czechoslovakia, 1

Dairy farming, 97-98, 101
Diets, 9-13

Ecuador, 79
Egypt, 18
Ethiopia, 74

Forage crops, 68-69
Fruits, 70-71, 96

Gathering, 87-88
Goats, 74, 77
Grain farming, 95-96, 101
Grazing, 21, 88

Hay, 68-69
Hogs, 72, 74

India, 1, 41, 59, 65, 67, 72, 74
Indonesia, 67
Iran, 18
Iraq, 18, 90
Ireland, 3
Italy, 1

Japan, 14-15, 65, 67
Jones, Wellington, 80

Kellogg, Charles E., 26-27
Keys, Ancel, 11

Labor, 82-83
Land
  capability, 29-31
  improvements, 83
  tenure, 82
  use, 20-26, 31, 82
  value, 82